Building the Future: A Guide to Semiconductors and Modern Electronics

Kiran

TABLE OF CONTENTS

Chapter 1 Introduction and Objectives

1.1 The rise of halide perovskites

In today's digital era, semiconductors are present in almost all modern electronics, including smartphones, computers, sensors, solar panels, memory storage, light-emitting diodes (LEDs), as shown in **Figure 1.1**. They have brought tremendous convenience and comfort to all human beings. The competition of coming fifth-generation (5G) communications and the internet of things (IoT) in these years are pushing the research of semiconductors and fabrications of electronics to an unprecedentedly charged field. Until now, the most widely used semiconductor material is still silicon, although abundant semiconductors have been discovered, synthesized, and utilized. So far, a large number of semiconductors have been synthesized in both elemental form (silicon, phosphorene, tellurene),[1-3] and compounds like III-V type gallium nitride (GaN),[4] II-VI type zinc oxide (ZnO),[5] IV-VI type silicon carbide (SiC),[6] I-II-VI$_2$ type copper indium sulfide (CuInS$_2$),[7, 8] layered materials like molybdenum sulfide (MoS$_2$),[9, 10] and organic materials.[11, 12] Advantages in synthesis, properties, and processing of novel semiconductor materials can always improve the performance of a wide range of electronic and optoelectronic devices, thus prompting the rapid development of new sciences and technologies for better human life. Therefore, the search of new semiconductors that can enable the exploration of superior properties and fabrication of better devices is one of the most critical goals and has long been pursued in materials science and engineering.

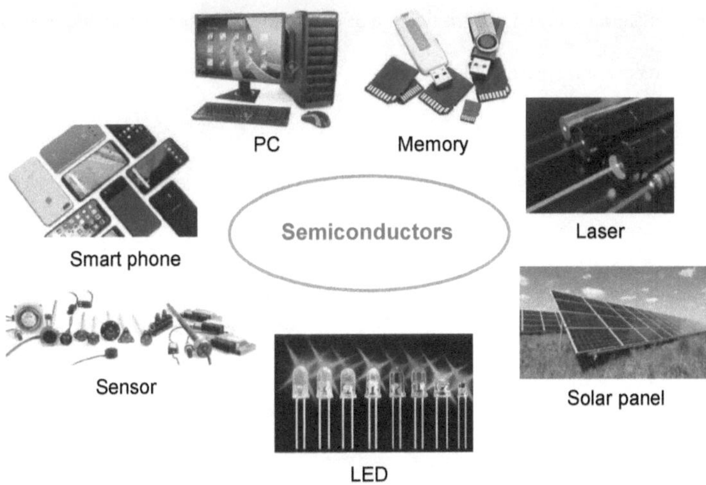

Figure 1.1 Electronics made of semiconductors like silicon in our daily life. Images of electronic devices are from Google.

The past few years have witnessed an explosive growth of interest in metal-halide perovskites since their first usage as visible-light sensitizers in photoelectrochemical cells in 2009 by Miyasaka with a power conversion efficiency (PCE) of 3.8%.[13] Due to efficient sunlight harvesting and charge transport the highest PCE of solar cells based on perovskites in labs has reached around 25% up to now in an all-solid-state structure.[14] Such a steep increase slope of the PCE increase seen from **Figure 1.2** has much exceeded other solar cells based on materials like silicon, cadmium telluride and copper-indium-gallium-diselenide.[15] This new kind of solar cells is believed to be able to compete with the commercial thin-film solar cells based on materials listed in Figure 1.2. Concurrent with the flourishing research on perovskite solar cells and thanks to the extensive efforts on material synthesis, the utilization of perovskites in photonic and optoelectronic devices has received increasing attention.

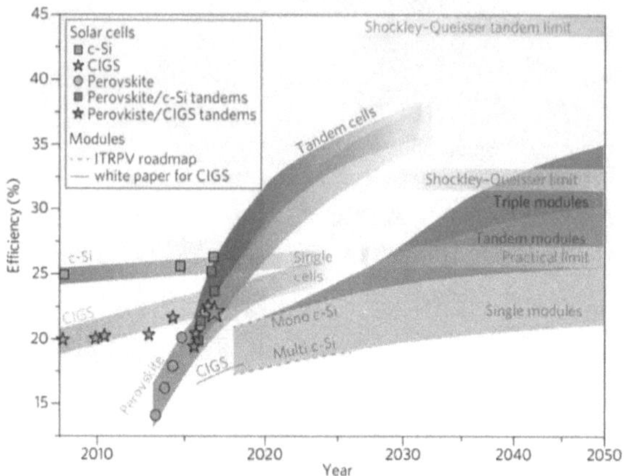

Figure 1.2 Achieved and predicted efficiencies for different solar cells. The purple ellipse indicates the PCE improvement of halide perovskite solar cells.[15]

Generally speaking, most halide perovskites can be categorized into three-dimension (3D) and two-dimension (2D) perovskites. The 3D perovskites have a formula of ABX_3, in which A represents either organic (methylammonium (MA) or formamidinium (FA): $CH(NH_2)_2^+$) or inorganic cations (Cs^+ or Rb^+), B is bivalent cations (Pb^{2+}, Sn^{2+}, Ge^{2+}, or Cu^{2+}) and X is halide anions (Cl^-, Br^-, or I^-).[16, 17] In this 3D structure, B cations and halide X anions constitute BX_6 octahedra, forming a continuous 3D skeleton, while A cations fill the spaces in between as shown in **Figure 1.3**a. The stability and lattice structure are determined by the tolerance factor t, which is defined as:

Equation 1.1 $$t = \frac{R_A + R_X}{\sqrt{2}(R_B + R_X)},$$

where R_A, R_B, and R_X are the radii of A, B, and X ions. The crystal with an ideal t value of 1 possesses a higher symmetry cubic structure. Deviations of t value from 1 usually give structural distortions and lattice structures with lower symmetries like orthorhombic and tetragonal phases. This structure is also featured with temperature-dependent phase transitions. For instance, the most studied perovskites, MAPbI$_3$, have a low-symmetry orthorhombic phase when the temperature is below 160 K and a tetragonal phase at a higher temperature, and it finally transforms into a cubic phase above 327 K as shown in **Figure 1.3b.**[18]

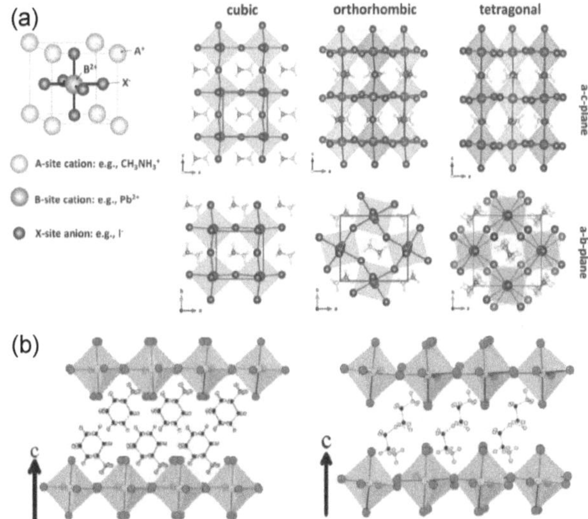

Figure 1.3 Halide perovskite schematic crystal structures. (a) 3D ABX$_3$ crystal structure. (b) MAPbI$_3$ perovskites with cubic, orthorhombic and tetragonal phases.[15] (c) (d) RP and DJ 2D perovskite crystal structures, respectively.[19]

In 2D layered perovskites, the large size of organic cations between the MX_6 layers determines their structures.[20] Depending on the ammonium ($-NH_3^+$) number in the organic cation, 2D layered perovskites can be assorted to Ruddlesden–Popper (RP) and Dion–Jacobson (DJ) types.[19, 21, 22] The RP perovskites have a formula of $(R-NH_3)_2BX_4$, in which R is the organic functional group in the cation and each organic cation has only one ammonium group, such as $C_4H_9NH_3^+$, phenylmethanamine ($C_6H_5CH_2NH_3^+$). In this structure, X atoms and organic layers between BX_6 layers are shifted between neighboring layers, which is a so-called "staggered" arrangement as shown in **Figure 1.3**c. The DJ perovskites have a formula of $(NH_3-R-NH_3)BX_4$, in which each organic cation has two ammonium groups, such as ethylenediammonium ($NH_3^+C_2H_4NH_3^+$)). In $(NH_3-R-NH_3)BX_4$ type, there is only one organic layer and the X atoms are aligned, giving rise to an "eclipsed" arrangement as shown in **Figure 1.3**d. Because the two organic layers are self-assembled through weak π-π interactions in 2D perovskites, atomic layer-thin flakes of 2D perovskites can be prepared by peeling from their bulk single crystals.

1.2   Halide perovskite synthesis, properties and applications

So far, plenty of methods have been invented to synthesize halide perovskites with various morphologies as shown in **Figure 1.4**. Polycrystalline films are the most studied ones. Low-temperature solution processing methods are the most common method to prepare polycrystalline halide perovskite films. Typical solution methods include "one-step" and "two-step" methods.[23] Take $CH_3NH_3PbI_3$ as an example. In the one-step method, MAI and PbI_2 are pre-mixed together in solvents and then spin-coated onto a substrate. To avoid pinholes and achieve a uniform film, usually, anti-solvents like chlorobenzene are dropped on the substrate near the end of spin-coating of halide perovskite solutions.[15] While in the two-step method, PbI_2 and MAI are spin-coated onto a substrate in a row. Then these two precursors react while annealing to form halide perovskite films. Beyond solution methods, vapor deposition or vapor solution combined methods can also be used to prepare films.[24, 25] For example, MAI and PbI_2 are evaporated simultaneously in a chamber. Their vapor reacts and cools down at the upper substrates as shown in **Figure 1.4**. The prepared film is also very smooth from the scanning electron microscopy image (SEM).[25] In addition, in vapor-included methods, PbI_2 films can be first prepared by solution or vapor methods, and then the PbI_2 films are converted to halide perovskite films.[24, 26] With the flourish developments of methods, it is convenient to prepare large-area, uniform, pinhole-free halide perovskite films.

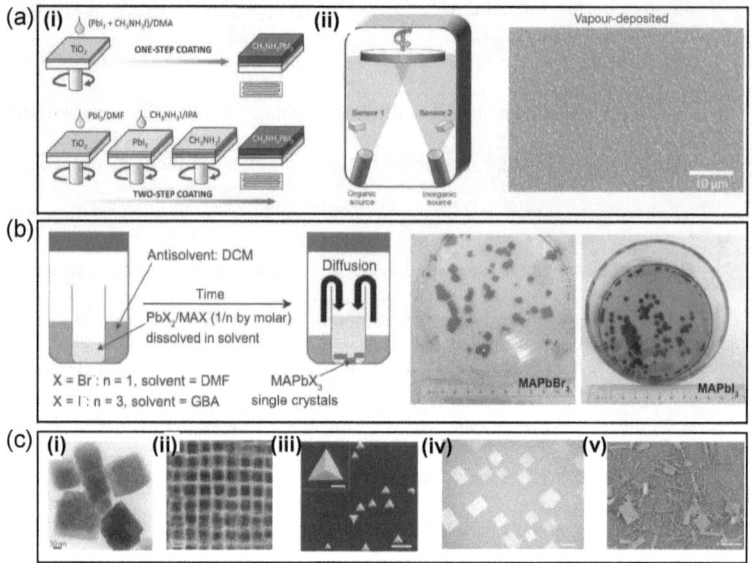

Figure 1.4 Different methods to prepare halide perovskite with different morphologies. (a) Polycrystalline films prepared by (i) one-step/two-step solution methods and (ii) Vapor method and the SEM image of halide perovskite film.[23, 25] (b) Anti-solvent vapor-assisted method to grow MAPbI₃ and MAPbBr₃ single crystals.[27] (c) Halide perovskites in small-scale with different morphologies. (i) Nanocrystals;[28] (ii) quantum dots;[28] (iii) pyramids;[29] (iv) nanoplatelets;[30] (v) nanowires.[31]

Besides polycrystalline films, single crystals and nano (micro) structures are also be able to be prepared as shown in **Figure 1.4**. The most employed approaches to growing halide perovskite single crystals are anti-solvent vapor-assisted and top-seeded solution growth methods. In the anti-solvent vapor-assisted method, vapors of anti-solvents like dichloromethane (DCM) gradually diffuse into the halide perovskite solutions, leading to the crystallization of halide perovskites.[27] In the top-seeded solution growth method, small single crystals are pre-synthesized by chemical reactions and used as seeds to grow larger single crystals in supersaturated halide perovskite solutions.[32] For MAPbI₃, the largest single crystal reported has a size of 71×54×39 mm.[33] Usually, the trap density in single crystals of halide perovskite is much less than their polycrystalline films. Taking MAPbI$_3$ as an example, the trap density is around 10^{10}-10^{11} cm^{-3} in the single crystal and 10^{15}-10^{16} cm^{-3} in the polycrystalline films.[34, 35]

Halide perovskites in small-scale have many morphologies, including nanocrystals,[28] quantum dots (QDs),[28] pyramids,[29] nanoplatelets,[30] and nanowires.[31] Solution and vapor methods are usually used to synthesize these morphologies. In solution, colloidal approaches with the presence of ligand are commonly used to prepare nanocrystals and QDs. The ligand is of great importance to control the shape, size, and growth mechanism.[36] In addition, by using lead precursors like lead acetate in a solid-state film, nanoplatelets and nanowires can be prepared with a reduced formation speed of halide perovskites due to the slow release of PbI$_4$$^{2-}$ from lead acetate. Besides self-grown, templets are also employed to shape the morphologies of halide perovskite from solution to crystallization.[37] Vapor method like chemical vapor deposition (CVD) is also widely to grow morphologies like pyramids,[29] nanoplatelets,[30, 38] and nanowires.[37]

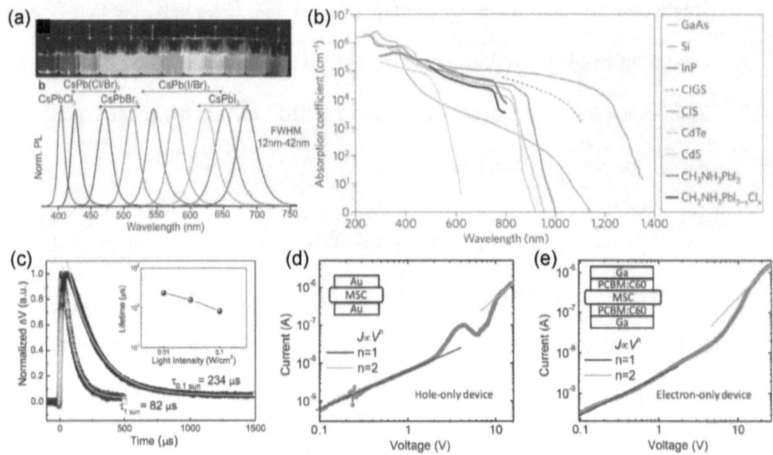

Figure 1.5 Physical properties of halide perovskites. (a) Tunable band gaps of $CsPbX_3$ with different halide compositions.[39] (b) Absorption coefficient of halide perovskites and other conventional semiconductors.[40] (c) Carrier life of $MAPbI_3$ single crystals under different light intensities.[32] (d) (e) Space-charge-limited current measurements of hole-only and electron-only $MAPbI_3$ devices, respectively.[13]

As a kind of new emerging materials, halide perovskites hold property advantages compared to typical semiconductors like silicon, III-V compounds like GaAs, InP, GaSb, and II-VI compounds like CdTe, CdS.[40] For example, the band gap of halide perovskites can be controlled from near-ultraviolet (around 3 eV for $MAPbCl_3$)[41-43] to near-infrared (NIR) region (1.4 eV for $FASnI_3$)[44, 45] continuously by modifying the composition. **Figure 1.5** present the photoluminescence (PL) spectra of $CsPbX_3$ with different halide compositions.[39] Apart from the halide, the metal B and cation A also play an important role in the band gap. For instance, the replacing of MA by FA and Pb by Sn can both decrease the band gap.[46] The photoluminescence quantum yield (PLQY) of halide perovskite nanocrystals can reach 90% for perovskite nanoparticles and 30% for films, which is much large than conventional semiconductor nanomaterials.[47] In addition, halide perovskites are strong light-harvesting materials with absorption coefficients higher than 10^4 cm^{-1}, which is much larger than silicon and comparable to binary compounds like CdTe, GaAs, and InP as shown in **Figure 1.5**.[40, 48] In terms of electrical transport properties, the carrier life time and carrier diffusion length are both quite high. In MAPbI$_3$ single crystals, the lifetime is 82 μs under 1 sun measured by a transient photovoltaic method, corresponding to a diffusion length of 175 mm.[49] The diffusion length is even larger at a lower illumination as shown in **Figure 1.5**. The carrier mobility of halide perovskite is also appealing. In MAPbI$_3$ single crystals, it is measured to be as high as 164 cm^2/Vs using the space-charge-limited current method (SCLC) or 105 cm^2/Vs using the Hall effect method as shown in **Figure 1.5**.[50]

The feasible preparation, excellent properties, and various morphologies make halide perovskites wide applications in photonic and optoelectronic devices. In reality, the applications of metal-halide perovskites in electronic and optoelectronic devices did not obtain obvious attention after the first synthesis. In 1958, Møller first reported that CsPbX$_3$ (X= Cl, Br, I) compounds had a perovskite structure, revealing the discovery of metal-halide perovskites.[51] Twenty years later, by replacing cesium with MA cation, Weber et al. reported the first organic-inorganic 3D hybrid metal halide perovskites.[52, 53] Since then, the cation in perovskites is not limited in inorganic ones, which widely extends the family members of halide perovskites. In 1994, Mitzi et al. reported a layered organic-based halide perovskite structure, in which a large organic cation, n-butylammonium ($C_4H_9NH_3^+$), was used, indicating the discovery of two-dimensional (2D) halide perovskites.[54] In 1999, the first thin-film field-effect transistor was fabricated using halide perovskites, initiating the applications of halide perovskites in electronics and optoelectronics.[55] In 2009, the application of metal-halide as light-harvesting materials in solar cells ignited radical research enthusiasm of metal-halide perovskites. After that, the research of electronic and optoelectronic devices including solar cells,[56] LEDs,[57] photodiodes/photoconductors,[26, 58] light-emitting transistors/phototransistors[59, 60], resistive memory and ferroelectrics are getting hotter and hotter.[61, 62] In more recent years, many new types of devices based on metal-halide perovskites are emerging, such as artificial synapses, photo ferroelectrics, and photo memory devices.[63-65]

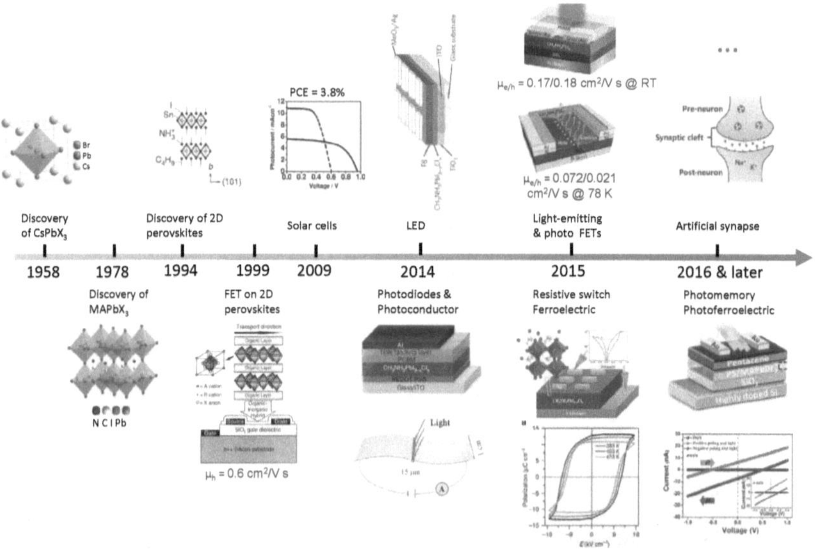

Figure 1.6 History of metal-halide perovskite synthesis and applications in electronic and optoelectronic devices.

1.3 Objective and structure of the book

As we can see above, metal-halide perovskites reveal great potential in electronic and optoelectronic devices because of their excellent chemical/physical properties and feasible preparation. Naturally, some questions can be proposed. Firstly, how can we improve the performance of these known devices? Take metal-halide perovskite solar cells as an example. The PCE of the first reported solar cell with metal-halide perovskites as light-harvesting materials is only 3.8%. Up to now, the PCE has jumped to around 26%.[14] However, the Shockley-Queisser limit of a single-junction perovskite solar cell is 33%,[15] leaving some space for chasing. A similar room for improvement also exists in photodetectors and other devices. Secondly, can we make new designs for new devices with specific functionalities? Take the operation band of metal-halide perovskite photodetectors as an example. The absorption of perovskites is limited to visible range (some can go a little bit far), although showing broad band gap tunability. Therefore, it is demanded to prepared devices with new architectures for IR detection.

We need to understand the origins of these problems to find solutions. Firstly, the most popular approach to prepare perovskite films is spin coating. In this way, the resultant film is poly-crystalline. The imperfections, including grain boundaries, high-density defects, and traps, bring extrinsic influences on the property investigations and detrimental effects on the device performance. For example, the imperfections significantly increase the non-radiative recombination probability in the perovskites, leading to low quantum efficiencies. In addition, the imperfections at the interfaces can impede the transfer of charges, causing substantial hysteresis and reduced charge extraction abilities.[66] Secondly, there are self-limitations in some devices or material structures. For example, in poly-crystalline perovskite planar photodetectors, a larger photoresponsivity needs a larger photocurrent, which requires the single-crystalline film. However, it also improves the dark current, which gives rise to a low photo detectivity.[67] Therefore, a trade-off between responsivity and detectivity happens. Thirdly, although possessing many excellent physical and chemical properties, metal-halide perovskites lack or show poor properties in smaller band gaps (IR), ferroelectrics, and chirality. Thus, perovskites are unlikely to be applied to devices with specific functionalities requiring these properties.

In the history of material research, hybridization with other materials or utilizing suitable substrates is a good way to modify the growth of high-quality films.[68-70] The involving of extrinsic materials during the growth can also be used to fabricate devices with new structures and novel functionalities.[69, 70] Inspired by these cases, a possible solution for the issues in metal-halide perovskite is to build heterostructures with other materials. Indeed, Numerous researches have demonstrated that heterostructures have wide applications in gate tunable, tunneling, thermionic, and light-harvesting and detecting devices.[71, 72] Among the large number of materials that can form heterostructures, 2D materials are unique considering their atomic-thin sheets and large surface-to-volume ratio. 2D materials can be metallic, semiconducting, and insulating, represented by graphene, black phosphorus/transition metal dichalcogenides (TMDCs), and hexagonal boron nitride, as shown in **Figure 1.7**, respectively. These 2D materials have to be combined with other zero-dimensional (0D), 1D, one-point-five-dimensional (1.5D) and 3D materials to form mixed dimensional 2D-0D, 2D-1D, 2D-1.5D, and 2D-3D heterostructures as shown in

Figure 1.7, providing unique properties and potential new functionalities compared to traditional semiconductor devices.[71, 72]

There could be several benefits from the heterostructures of metal-halide perovskites with other materials. First, the extrinsic materials could adjust the growth of perovskites, whether in vapor growth as substrates or in solution as additives. The dangling bond-free surface of 2D materials may lead to different growth modes like van der Waals (vdW) epitaxy of perovskites, which has been verified in other cases like CdS, GaN, and GaSe.[73-75] Second, the properties of perovskites and the extrinsic materials can be combined together, remedying the drawbacks of perovskites. A clear case is the absorption in NIR/IR regions can be compensated by other materials.[76, 77] Third, new properties are possible to be proposed due to the synergetic effects between two different materials. A type-II heterojunction formed is likely to generate faster charge separation and transfer.[78] At last, new devices with novel structures and functionalities may be invented, which has been proved obviously in the 2D van der Waals heterostructures.[71] Therefore, it is reasonable to speculate that the performance and functionalities of halide perovskite devices can be improved and broadened by establishing heterostructures with 2D materials.

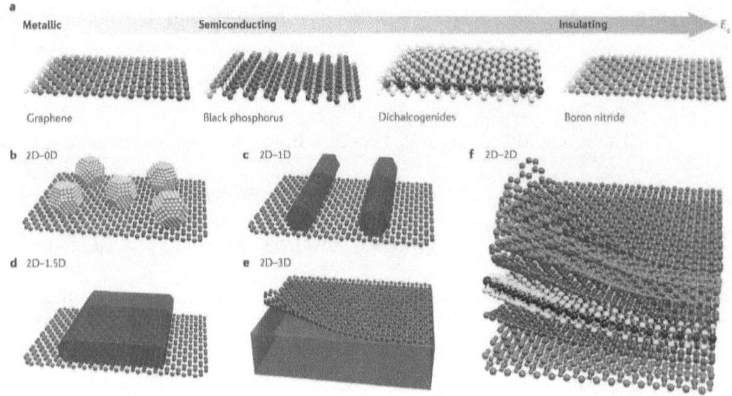

Figure 1.7 Different types of 2D materials forming different dimensional heterostructures. The 2D materials can be from metallic to semiconducting and to insulating, while the other materials can be 0D, 1D, 2D, 1.5D, and 3D.[71]

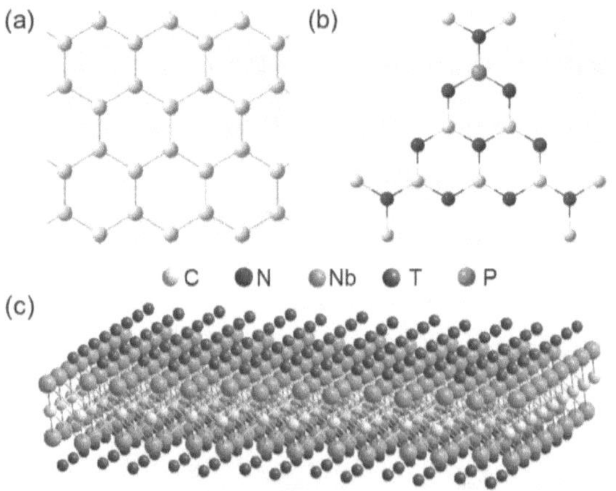

Figure 1.8 2D materials applied to prepared heterostructures with metal-halide perovskites (a) Graphene. (b) Phosphorus doped g-C_3N_4. (c) Nb_2CT_x MXene.

Since a large number of 2D materials have been reported, it is importance to select the most suitable 2D materials. Halide perovskite heterojunctions are formed with different kinds of 2D materials, from metallic to semiconducting. Graphene is chosen since it is a very common semimetallic 2D material. In addition, graphene can provide a large single-crystalline area as a substrate for the growth single-crystalline halide perovskites. Therefore, it offers a platform to investigate their interactions without extrinsic effects from grain boundaries and defects. Some interactions between halide perovskites and 2D materials are general while some are certainly not. Therefore, besides the semi-metallic graphene, the semiconducting graphitic carbon nitride (g-C_3N_4) is selected as another semiconducting 2D material. The synthesis of g-C_3N_4 and phosphorus-doped g-C_3N_4 is cheap and convenient, demanding no subtle conditions like other 2D materials. The band gap and electronic structure of g-C_3N_4 can also be tuned by doping of phosphorus, imbuing more tunability of heterostructure properties and the resulted device performance unlike graphene. Last, the band gap of halide perovskites are usually limited in UV-visible range as mentioned above. Heterostructures of halide perovskites and other 2D materials can expand the optoelectronic applications of halide perovksites beyond their absorption cutoff. A relatively new kind of 2D material called Nb_2CT_x MXenes is such a material, considering its unique plasmonic absorption feature in the NIR region. The lattice structure of these materials is shown in **Figure 1.8**.

In the remainder of this book, the following chapters are covered, which are briefly summarized below:

<u>Chapter 2</u>: We report the synthesis of a mixed-dimensional heterojunction composed of 3D MAPbBr$_3$ single-crystal platelets on 2D single-layer graphene. Although MAPbBr$_3$ has a nonlayered lattice structure, single-crystal platelets with an exclusive (001) orientation were selectively grown on single-layer graphene via van der Waals epitaxy using a one-step chemical vapor deposition (CVD) method. A complementary set of techniques, including PL, Raman spectroscopy, Kelvin probe force microscopy (KPFM) were used to unravel the charge transfer at the perovskite/graphene interface. The amount of hole doping in graphene was estimated to be 7.5×10^{12} cm^{-2}, which is accompanied by a Fermi level decrease of 272 meV. The charge transfer characteristics were further confirmed using a field-effect phototransistor with the perovskite/graphene heterostructure as the channel. This work is expected to enrich the vdW synthesis of 3D/2D mixed-dimensional heterostructures and to propel their optoelectronic applications.

<u>Chapter 3</u>: Efficient charge transfer is verified between 2D materials and perovskites in the above work. Such charge transfer can be utilized to improve the performance of devices like photodetectors. Meanwhile, 2D materials that can be synthesized in simple and commercial ways are favored for practical applications. Therefore, in this work, we fabricate photodetectors made of methylammonium lead tri-halide perovskite (MAPbI$_{3-x}$Cl$_x$: MLHP) and phosphorus-doped graphitic carbon nitride nanosheets (PCN-S). Using thermal polymerization, PCN-S with a reduced band gap, are synthesized using low-cost precursors, making it feasible to form type-II bulk heterojunctions with perovskites. Owing to the bulk heterojunctions between PCN-S and MLHP, the dark current of the photodetectors significantly decreases from $\sim 10^{-9}$ A for

perovskite-only devices to ~10^{-11} A for heterojunction devices. As a result, not only the on/off ratio of the hybrid devices increases from 10^3 to 10^5, but also the photodetectivity is enhanced by more than one order of magnitude (up to 10^{13} *Jones*), and the responsivity reaches a value of 14 A W^{-1}. Moreover, the hybridization of MLHP with PCN-S significantly modifies the hydrophilicity and morphology of the perovskite films, which dramatically increases their stability under ambient conditions. The hybrid photodetectors, described here, present a promising new direction towards stable and efficient optoelectronic applications.

Chapter 4: So perovskite photodetectors with improved and stable performance are prepared. However, the operation band of most perovskite photodetectors is still limited in the UV-visible range. Therefore, it is essential to broaden the detection band of perovskite photodetectors. In previous work, MXenes have shown impressive plasmonic absorptions spanning the visible and infrared (IR) regimes. However, their potential IR optoelectronic applications including photodetectors, are marginally investigated and only limited to $Ti_3C_2T_x$ MXene. Besides, their relatively low resistivity restricts their use as photo-sensing materials due to the natural large dark current. In this work, heterostructures made of methylammonium lead tri-halide perovskite (MAPbI$_3$) and Nb_2CT_x MXene with a matching band structure are successfully prepared and exploited for self-powered visible-near infrared (NIR) photodiodes. The use of the MAPbI$_3$ layer expands the operation of the diode to the visible range while suppressing the dark current of the NIR-absorbing Nb_2CT_x layer. As a result, the photodiode responds linearly under white light illumination with a responsivity of 0.25 A/W. Furthermore, when illuminated with a 1064-nm laser, the photodiode demonstrates a higher on/off ratio ($\sim10^2$) and faster response times (< 30 ms) compared to that of planar Nb_2CT_x-only detectors (< 2 and 20 s, respectively). Experiments from space-charge limited current and capacitance measurements show that the coordinate bonding between the surface groups of the MXene and the undercoordinated Pb^{2+} ions in the $MAPbI_3$ surface has led to a passivated $MAPbI_3/Nb_2CT_x$ interface resulting in an efficient and speeded charge transfer.

Chapter 5: The research achievements are summarized. Also, we discuss several potential research directions of perovskite-2D materials heterostructures, including growth, patterning, integration, and more types of heterostructures and devices of high-quality perovskite films.

Chapter 2 Single-Crystal Hybrid Perovskite Platelets on Graphene: a Mixed-Dimensional van der Waals Heterostructure

2.1 Summary

Van der Waals (vdW) heterostructures open up excellent prospects in electronic and optoelectronic applications. It is of great importance to prepare high-quality like single-crystalline heterostructures to investigate the interaction such as charge transfer on the interface. Therefore, mixed-dimensional metal-halide perovskite/graphene heterostructures are prepared through selective growth of $MAPbBr_3$ platelets on patterned monolayer graphene using the CVD method. Preferred growth of single-crystal $MAPbBr_3$ platelets on graphene surfaces is achieved, which is accompanied by significant photoluminescence quenching. Raman spectra reveal that perovskite platelets cause p-type doping in the graphene layer. A significant Fermi level decrease of 272 meV in graphene is estimated, which corresponds to a high doping density of 7.5×10^{12} cm^{-2}. Surface potentials measured by Kelvin probe force microscopy indicate a negatively charged perovskite surface under illumination, which is consistent with the upward band bending deduced from conducting atomic force microscopy measurements. Moreover, a field-effect phototransistor is fabricated using the perovskite/graphene heterostructure channel, and the increased Dirac voltage under illumination confirms an enhanced p-type character in graphene. These findings enrich the understanding of strong interface coupling in such mixed-dimensional vdW heterostructure and pave the way towards novel perovskite-based optoelectronic devices.

2.2 Introduction and background

VdW heterostructures, in which neighboring layers are weakly bonded by vdW interactions, have established a new platform for fundamental scientific researches and novel device applications.[71, 72] Synergetic combinations of 2D materials, including graphene and transition-metal dichalcogenides, have led to the construction of vdW heterostructures with diverse phenomena and properties. This emerging materials-assembling strategy opens up new avenues for building innovative electronic and optoelectronic devices such as tunneling transistors,[79] barristors,[80] gate-tunable diodes,[81] photodetectors,[82] and light-emitting devices.[83] Beyond 2D/2D stacks, mixed-dimensional (nD/2D or 2D/nD, where n = 1 or 3) vdW heterostructures are more difficult to synthesize due to non-passivated and subtle interfaces. Nevertheless, there have been some reports on mixed-dimensional vdW heterostructures, such as PbS/graphene,[84] PbS/MoS_2,[75] CdS/MoS_2,[74] MoS_2/GaN,[85] perovskite/WS_2[86] and MoS_2/Si[87] prepared by delicate high-temperature physical/chemical vapor deposition methods, which differ from stacking of 2D materials by mechanical methods. The contact areas of these heterojunctions are usually less than 1×1 μm^2 as a result of the non-passivated interface of at least of one material.[74, 75, 84, 85] These mixed-dimensional vdW heterostructures possess unique advantages such as faster charge transfer and tailored energy band alignments.[71, 78, 83, 88, 89] Thus, mixed-dimensional vdW structures with complementary materials and optimal properties are highly promising for potential applications in optics and electronics.

Metal-halide perovskites have received intense research efforts owing to their extraordinary optical and electrical properties.[15, 40] Besides solution-processed polycrystalline films, bulk crystals and nanocrystals, large-scale ultrathin perovskites via vapor-phase growth have attracted intensive research efforts.[37] It is hopeful that such thin perovskites may simultaneously possess the merits of bulk crystals and thin film, enabling devices with high performance. For example, polycrystalline $MAPbI_3$ platelets were grown on graphene, MoS_2 and h-BN to form vdW solids following a two-step method, and charge transfer was investigated in such heterostructures.[90] Also, $MAPbBr_3$ and $CsPbBr_3$ platelets were grown directly on 2D mica substrates via vdW epitaxy.[38, 91] These pioneering works demonstrate a promising platform of perovskite-including vdW heterojunctions for future optoelectronic and photovoltaic applications. However, controlled growth of ultrathin perovskite-based single-crystalline vdW heterostructures and thorough investigation of their physical properties, particularly those related to the hidden interfaces, remain a highly challenging issue.

2.3   Experiment section

2.3.1   Material preparation

Graphene transfer: Monolayer graphene on copper foils was purchased from a commercial company (UniversityWafer, Inc.). To transfer the graphene from copper foils to SiO_2/Si substrates, a poly(methyl methacrylate) (PMMA) layer was first spin-coated onto the graphene film on a copper foil. After annealing for 20 min at 150 °C, the copper foil with PMMA was put into copper enchant (Sigma-Aldrich). After the copper was fully removed, the PMMA/graphene was floated on the surface and moved to distilled water. After being rinsed several times to remove the enchant residue, the PMMA/graphene was carefully moved to a SiO_2/Si substrate beneath. Then the SiO_2/Si substrate with PMMA/graphene was annealed at 180 °C for 30 min to flatten the graphene film on the substrate. Last, the PMMA on the graphene was removed by washing with acetone for enough times.

$MAPbBr_3$ platelet growth: $MAPbBr_3$ platelets were prepared using a one-step CVD method from lead bromide ($PbBr_2$) and methylammonium bromide (MABr) powders. Graphene transferred on SiO_2/Si substrates were used as the substrates for growth. For selective growth, the graphene was patterned by photolithography and then etched by O_2 plasma for 15 s at a power of 60 W. $PbBr_2$ powder was put in the center of the furnace (325 Torr) while MABr and substrates were put in the up- and downstream zones, respectively. The process was carried at a pressure of 100 torr under the flow of N_2 with a speed of 30 sccm.

2.3.2 Phototransistor fabrication

For phototransistor fabrications, 60 nm Au was deposited using e-beam evaporation as source and drain electrodes with a shadow mask. The bare graphene transistors were fabricated on graphene without been patterned.

2.3.3 Characterizations

Optical microscope (OM) images were taken from Nikon ECLIPSE LV100N POL. X-ray diffraction (XRD) patterns were recorded from Bruker diffractometer (D8 Advance) using Cu Kα (λ=1.5406 Å) radiation. μ-Raman and photoluminescence (PL) spectra were obtained using a Hariba LabRAM HR spectrometer with a He–Ne laser (spot size 1 μm through an objective lens) with a wavelength of 473 nm. The intensity was low enough to avoid damage. Kelvin probe force microscopy (KPFM) and conducting atomic force microscopy (CAFM) were measured on a commercial AFM (Asylum Research MFP-3D) using a stiff Pt/Ir-coated silicon tip with a spring constant of 40 N/m. Current-voltage (*I-V)* curves of the phototransistors were measured using a Keysight B1500A semiconductor device parameter analyzer in ambient conditions.

2.4   Results and discussions

2.4.1   Growth of MAPbBr₃ platelets on graphene

Successful transfer of monolayer graphene from copper foils to SiO_2/Si substrates is demonstrated by the AFM images in **Figure 2.1**. The graphene was scratched to measure the thickness. A thickness of around 0.4 nm reveals the graphene is monolayered indeed. After transfer, no obvious residue is observed on the graphene surface, which is qualified for the growth of MAPbBr₃.

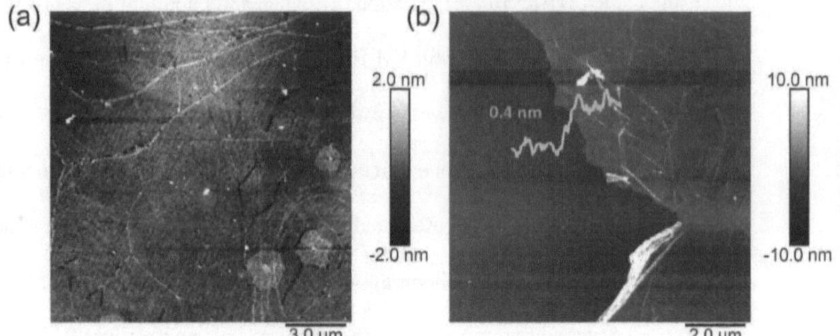

Figure 2.1 Transfer of monolayer graphene from copper foils to SiO_2/Si substrates. (a) Monolayer graphene on a copper foil. (b) Monolayer graphene transferred to SiO_2/Si substrates.

The MAPbBr$_3$ platelets were grown on monolayer graphene using a one-step CVD method as illustrated in **Figure 2.2**a. MAPbBr$_3$ platelets were grown from vapors produced by lead bromide (PbBr$_2$) and methylammonium bromide (MABr) powder precursors. **Figure 2.2**b shows the lattices of MAPbBr$_3$ and graphene and the growth of MAPbBr$_3$ platelets on monolayer graphene. To the best of our knowledge, it is the first time that single-crystal hybrid perovskites are demonstrated to grow on the graphene surface via a one-step method. It should be noted that the two-step growth of hybrid perovskite platelets on graphene has been reported.[90] However, the two-step conversion process can not guarantee the single-crystalline nature of perovskite platelets, and optoelectronic devices based on such heterostructures have not yet been demonstrated. **Figure 2.2**c-d shows the OM images at different magnifications of some perovskite platelets grown on graphene/SiO_2/Si substrates. It can be seen that MAPbBr$_3$ nucleates randomly on the graphene surface and grows fast laterally due to the small migration free energy barrier along the graphene surface.[38, 92] In the conventional epitaxial growth, a lattice match at the interface is a prerequisite. However, in our case, the lattice constant of cubic MAPbBr$_3$ (5.92 Å) does not match that of honeycomb graphene (2.46 Å) at all.[38, 93] Therefore, a weak van der Waals interaction that demands no lattice match between perovskite and graphene must be responsible for the growth. This kind of van der Waals epitaxy has been widely reported in several cases of 2D/2D and 3D/2D material stacks.[30, 89]

Figure 2.3a-b show the AFM topography and phase images of a MAPbBr$_3$ platelet. The square-shaped platelet has a thickness of 21 nm and a lateral size of 3 μm. Interestingly, a smaller platelet was found in the phase image, but it is invisible in the topography image due to its much smaller thickness, and its shape is also less defined than the larger one, indicating an early growth stage of the platelet. The powder XRD pattern in **Figure 2.3**c shows a set of sharp diffraction peaks that can be assigned to the cubic MAPbBr$_3$ phase. There are some weak peaks not in the (00l) orientations, which may be caused by the existence of some perovskite particles with irregular shapes on the graphene surface. The PL spectrum in **Figure 2.3**d further confirms that the as-prepared platelets are MAPbBr$_3$ with a band gap of 2.3 eV. However, the peak intensity of the platelets on graphene is much reduced compared to the platelet on insulating muscovite mica substrates, indicating efficient charge transfer between MAPbBr$_3$ platelets and graphene.

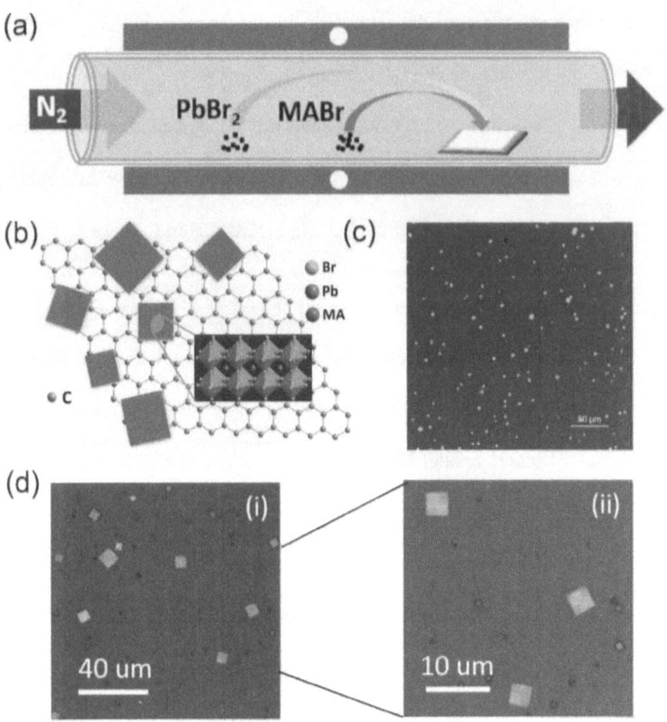

Figure 2.2 CVD growth of MAPbBr₃ on monolayer graphene. (a) Schematic of the CVD growth process. (b) Schematic of MAPBBr₃ lattice and their platelets on monolayer graphene. (c) (d) MAPbBr₃ platelets grown on monolayer graphene with different scales.

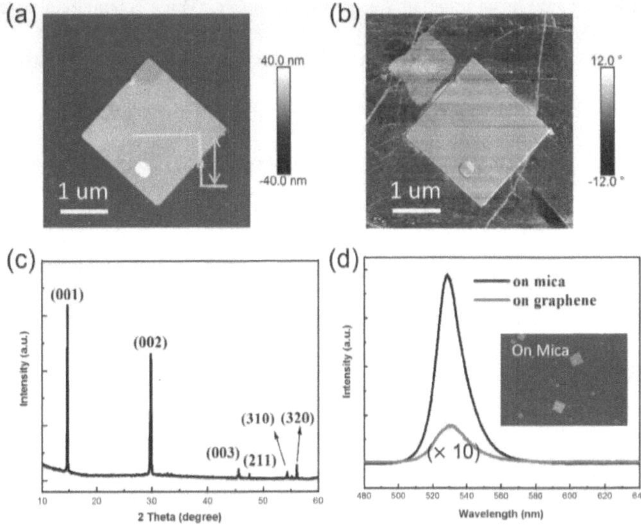

Figure 2.3 Characterizations of the MAPbBr₃ platelets on monolayer graphene. (a)AFM topography and (b) phase images of MAPbBr₃ platelets on graphene. (c) XRD spectrum of the as-grown MAPbBr₃ platelets on graphene. (d) PL spectra of MAPbBr₃ grown on graphene and mica substrates. Inset is an OM image of MAPbBr₃ platelets grown on mica.

2.4.2 Selective growth of MAPbBr₃ platelets on monolayer graphene

To investigate the effect of graphene on the growth of MAPbBr₃ platelets, SiO_2/Si substrates partially covered with graphene were used as shown in **Figure 2.4**a. OM images in **Figure 2.4**b-d reveal the growth of MAPbBr₃ at three different locations along the graphene edge with a length of 1.5 cm on the substrate. The temperature (T_L) at the location of **Figure 2.4**b is approximately 10 °C lower than the temperature (T_H) at a downstream location of **Figure 2.4**d. Clearly, MAPbBr₃ platelets are more likely to grow on the edge of the graphene, which mirrors the previous reports that PbS is likely to grow along edges of graphene ribbons and MoS_2 sheets.[8,9] This edge effect can be ascribed to the fact that the edges with unsaturated atoms are highly active, which is beneficial to the nucleation of perovskite platelets. Furthermore, with the increase of temperature, the number and lateral size of the platelets increase, owing to the thermally accelerated nucleation and growth speed.[38] As a result, the density of platelets on the graphene interior as shown in **Figure 2.4**e becomes comparable to that along the edge at high temperatures. However, on the SiO_2 side, under the same synthetic conditions, only particles with irregular shapes can be observed as a result of a high energy barrier for nucleation and lateral growth as shown in **Figure 2.4**f.[38, 92]

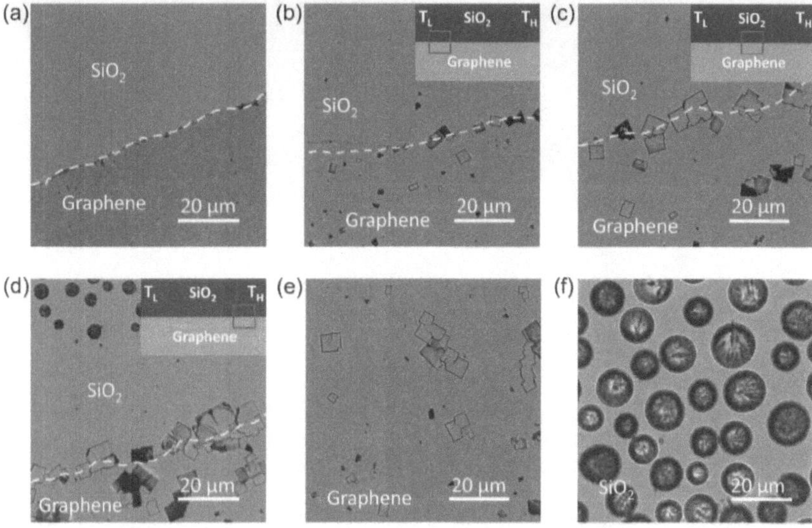

Figure 2.4 Selective growth of single-crystal MAPbBr$_3$ platelets on graphene. (a) Optical image of a SiO$_2$/Si substrate partially covered with monolayer graphene. The white dash line indicates the graphene edge. (b-d) Morphologies of MAPbBr$_3$ platelets grown on SiO$_2$/Si substrates at three different temperature zones. The insets illustrate the locations on the substrate. T$_L$ and T$_H$ represent low-temperature and high-temperature zones in the furnace, respectively. (e) (f) MAPbBr$_3$ grown on graphene interior and on SiO$_2$/Si, respectively.

To further verify the selective growth, graphene was patterned to strips on SiO$_2$/Si substrates by photolithography before growth. As shown in **Figure 2.5**, MAPbBr$_3$ platelets tend to grow on the graphene stripes with good selectivity and to contact the graphene edges, which strongly demonstrate the selective growth of MAPbBr$_3$.

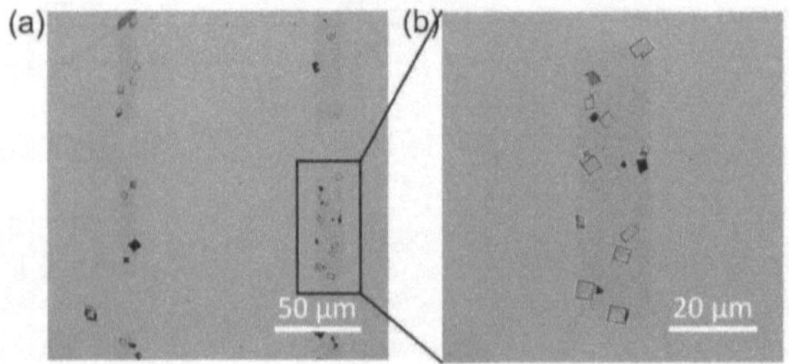

Figure 2.5 MAPbBr$_3$ platelets grown on patterned graphene strips.

2.4.3 Raman spectrum

The interaction between graphene and MAPbBr$_3$ was further investigated using µ-Raman spectroscopy as shown in **Figure 2.6**a-c. Raman shifts of graphene with and without the growth of MAPbBr$_3$ platelets were recorded at three different locations to exclude the influence of any inhomogeneity. As reported in previous studies, the G band and 2D band of monolayer graphene are located at around 1580 cm^{-1} and 2700 cm^{-1}, respectively.[94] As shown in Figure 3b, the G band appears at 1582.8 cm^{-1} and 1595.7 cm^{-1} for graphene and MAPbBr$_3$-covered graphene, respectively. For the 2D band, these values are 2700.3 cm^{-1}, and 2731.3 cm^{-1}, respectively. It is worth noting that only a small bump is observed around 1350 cm^{-1}, which is the D band caused by localized defects, indicating the high quality of the transferred graphene.[90] During Raman measurements, the intensity of the laser was low enough, and no damage was observed on the sample as shown in **Figure 2.6**d.

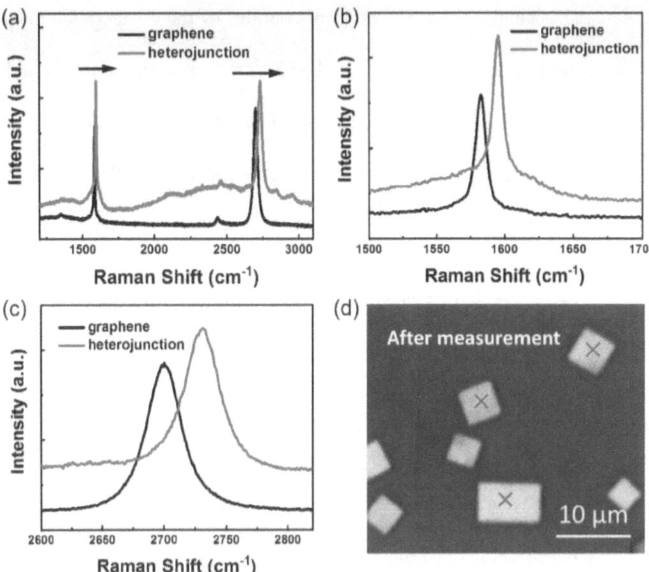

Figure 2.6 Raman measurements of MAPbBr$_3$/graphene heterostructures. (a) μ-Raman spectra of graphene with and without the growth of MAPbBr$_3$ platelets. (b), (c) Close-up views of G and 2D peaks, respectively. (d) OM images of graphene covered with MAPbBr$_3$ platelets

Since G and 2D band positions are sensitively dependent on the carrier concentration, they can be used to probe the graphene doping level.[73, 94, 95] Clearly, compared to the uncovered graphene, the covered one displays blue-shifted G and 2D bands. The blue-shifted 2D band indicates *p*-type doping to graphene after the growth of MAPbBr$_3$ platelets.[73, 95] Furthermore, the Fermi level of graphene (E_F) can be derived from the shift of the G band wavenumber (ω_G) based on the formula[73, 95]

Equation 2.1 $\quad\quad \omega_G - 1580 = |E_F| \times 42 \text{ cm}^{-1} \text{ eV}^{-1}.$

Accordingly, the average Fermi energies of the covered and uncovered graphene were calculated to be -90 meV and -362 meV, respectively. Thus, interfacing with perovskite causes a Fermi level decrease of 272 meV for graphene. This change is much larger than the values reported for C_{60}/graphene (170 meV) and GaSe/graphene (80 meV),[73, 95] demonstrating a stronger interaction between graphene and MAPbBr$_3$ platelets.

The carrier density can be estimated from the corresponding Fermi energies. The relationship between the carrier density N and the Fermi level E_F of graphene is

Equation 2.2 $\qquad |E_F| = \hbar\, v_F\, (\pi\, |N|)^{1/2},$

where v_F is the Fermi velocity of the linear band and $\hbar$ is the reduced Planck constant.[73] Using a Fermi velocity of $v_F = 1.1 \times 10^6$ m/s, the hole densities for graphene with and without the growth of MAPbBr$_3$ platelets are 4.9×10^{11} cm^{-2} and 8.0×10^{12} cm^{-2}, respectively. Therefore, the doping density from MAPbBr$_3$ platelets to graphene is 7.5×10^{12} cm^{-2}, which is larger than the values reported on heterostructures of C_{60}/graphene (5.7×10^{12} cm^{-2}) and GaSe/graphene (5×10^{12} cm^{-2}).[73, 95] It is worth noting that strain in graphene can also lead to Raman shifts.[96] However, this effect can be excluded because the van der Waals growth should not bring much strain. Also, the strain would cause the G band to red-shift, which is contradictory to our results.[96]

2.4.4 Surface potential

The interaction between MAPbBr$_3$ platelets and graphene was further investigated using KPFM. The surface potential reflected by the contact potential difference (V_{CPD}) can offer important insights on the localized band structures.[39-41] Typical AFM topography of the platelets is shown in **Figure 2.7**a, and the thicknesses of the platelets are 20 nm, 23 nm and 37 nm, respectively. As shown in **Figure 2.7**b-c, the V_{CPD} of graphene does not change much, while the V_{CPD} of MAPbBr$_3$ platelets under light is notably lower than that in the dark, suggesting more negatively charged surfaces.[40] The single-crystal characteristic of the platelets can eliminate parasitic effects brought by grain boundaries, including accumulated ions, defects, and traps.[39] Therefore, the ion accumulation can be excluded in our KPFM measurements, and the negative charges should be attributed to the electron accumulation on the platelet surfaces.

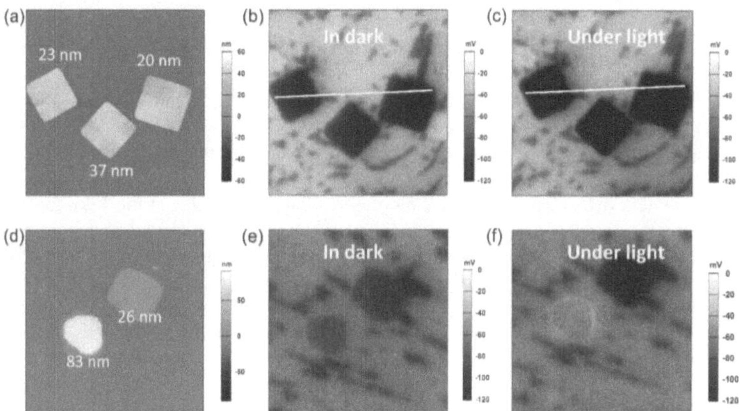

Figure 2.7 KPFM measurements of MAPbBr$_3$ platelets. (a) AFM topographic image of MAPbBr$_3$ platelets with thicknesses of 20 nm, 23 nm and 37 nm. (b) (c) KPFM images under dark and white light (intensity ~ 1 mW/cm^2) conditions. The lines indicate the positions of the V_{CPD} profiler. (d-f) Same measurements as (a-c) for platelets with thicknesses of 26 nm and 83 nm.

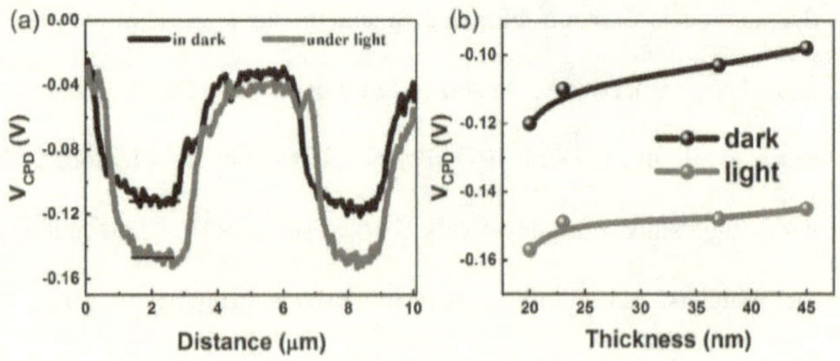

Figure 2.8 (a) Line profiles of V_{CPD} along white lines in Figure 2.7 (b) and (c) under dark and light conditions. (h) V_{CPD} under dark and light conditions for platelets with different thicknesses.

KPFM measurements were also carried out on other MAPbBr$_3$ platelets from another sample with thicknesses of 26 nm and 83 nm in **Figure 2.7**d-f. Interestingly, the V_{CPD} for the 83-nm-thick platelet becomes much higher and is almost the same as the graphene nearby, which is very different from the thinner platelets and indicates a strong thickness dependence. **Figure 2.8**a presents the line profiles of the V_{CPD} for platelets with thicknesses of 23 nm and 20 nm under dark and light illumination conditions. Around a 40 meV drop of V_{CPD} is observed from dark to light conditions. The V_{CPD} values measured on platelets with four different thicknesses are summarized in **Figure 2.8**b. It should be noted that only V_{CPD} measured on the same substrate and roughly at the same time can be compared since environmental changes have a huge impact on KPFM results.[42] As a result, the quantitative comparison of V_{CPD} between data presented in **Figure 2.7**a-c and those in **Figure 2.7**d-f is not valid since they were taken on different samples under different ambient conditions. Nevertheless, the consistent increase of V_{CPD} indicates that electrons accumulate in the MAPbBr$_3$ platelet surfaces under light illumination, while holes transfer to the underneath graphene. Thus, the KPFM data are consistent with the scenario of light-induced p-type doping to graphene and upward bent bands in the space charge region of MAPbBr$_3$.

2.4.5   Charge transportation characteristic

The band bending was also deduced from the *I-V* curves measured using CAFM as shown in **Figure 2.9**. The height image of the measured sample and the marks of measured locations are also shown. The inset shows a schematic illustration of the experiment setup, in which a Pt-Ir coated tip is employed to apply voltages. A quasi-linear *I-V* curve was obtained for graphene, which is expected because the Pt-Ir CAFM tip has a very large work function of approximately 5.1 eV. In contrast, a typical diode behavior with a rectifying *I-V* curve was found for the heterojunction in the dark. It is forward-biased when a negative voltage is applied to the tip. Such a rectifying characteristic implies that a Schottky barrier exists when holes are injected from p-type graphene to MAPbBr$_3$, which is consistent with the upward bent band edges of MAPbBr$_3$ at the MAPbBr$_3$/graphene interface. Under light illumination, the reverse-biased current is significantly enhanced due to the photo-generated carriers,[97] but the rectification is retained. Such a drastic light-induced modification of the transport behavior indicates that such mixed-dimensional vdW heterostructures are promising for optoelectronic applications.

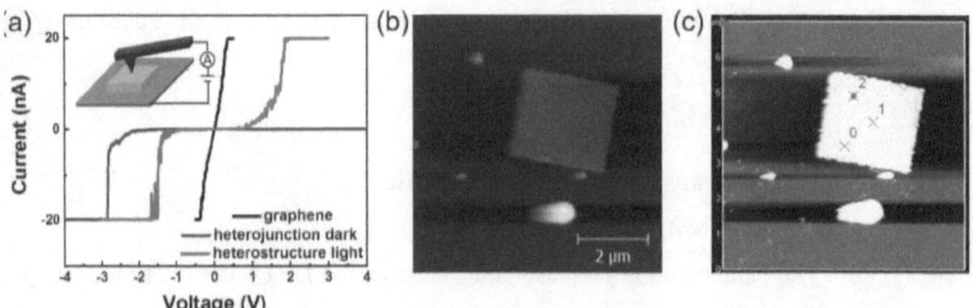

Figure 2.9 CAFM measurements. (a) *I-V* curves of graphene and a MAPbBr$_3$/graphene heterojunction under dark and light conditions. The inset is a schematic of the CAFM setup, in which the grey bottom plate is graphene, the orange thin slab is MAPbBr$_3$ platelet, and the dark blue pyramid is the CAFM tip. Field-effect phototransistors. (b) Height image of the MAPbBr$_3$ platelet for CAFM measurement. (c) Marks show the tip positions for measurements.

2.4.6 Field-effect phototransistors

To further investigate the interfacial interaction, field-effect transistors were fabricated based on the graphene and MAPbBr$_3$/graphene heterostructures. The typical device has a channel length of 30 μm and a width of 10 μm. Transfer curves of the as-fabricated devices are shown in **Figure 2.10**a. The on/off ratio of the graphene-only device is approximately 3.5, which is comparable to previous reports.[98, 99] The positive Dirac voltage (V_{Dirac}) of 60 V indicates a p-type character of the pristine graphene,[99] which is in agreement with the Raman shift. The photoresponse of the pristine graphene is negligible in the white light illumination condition, as shown in **Figure 2.10**b. For the heterojunction-based devices, upon light illumination, both source-drain current (I_{ds}) and V_{Dirac} increase, indicating that holes are transferred to graphene with the presence of MAPbBr$_3$ platelets. **Figure 2.10**c shows the photo-response of the perovskite/graphene channel under zero gate voltage. The inset is the image of the device, and the coverage of MAPbBr$_3$ platelets on the graphene channel is approximately 20%. An 11% current enhancement under a 0.1 V source-drain bias is deduced, revealing a large amount of hole doping from the MAPbBr$_3$ platelets.

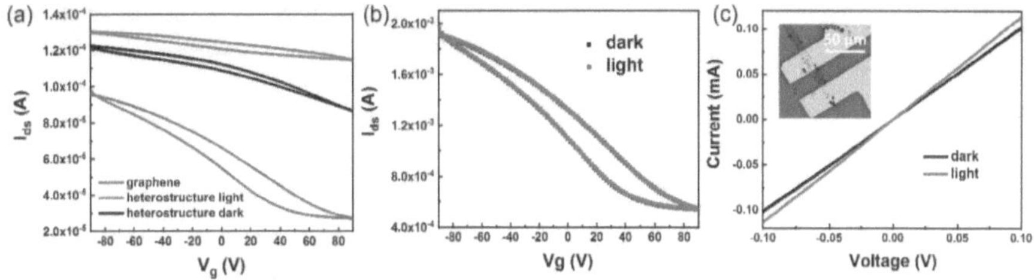

Figure 2.10 MAPbBr$_3$/graphene phototransistor. (a) Transfer curves of graphene and MAPbBr$_3$/graphene field-effect transistors under dark and light illumination with a source-drain voltage of 0.1 V. (b) Photoresponse for a graphene field-effect transistor. (c) *I-V* curves of MAPbBr$_3$/graphene phototransistor under dark and white light conditions (intensity: 1 mW/cm^2). Inset shows the photo of the device.

2.4.7 Band diagram

Therefore, the aforementioned G and 2D peak shifts in Raman, V_{CPD} drops in KPFM, and the Dirac voltage increase in field-effect phototransistor consistently manifest the *p*-doping to single-layer graphene after the growth of MAPbBr$_3$ platelets. Given the work function of pristine graphene (4.5 eV),[35,36] and band structures of MAPbBr$_3$ (electron affinity: 3.7 eV, work function: 4.0 eV, and band gap of 2.3 eV, *n*-type),[46,47] charge transfer can be understood by determining the band offsets resulting from the junction formation. In the vdW epitaxial structure of MAPbBr$_3$/graphene, interface states should be negligible because both materials are single crystals, and the graphene surface is free of dangling bonds.[48] The weak vdW interactions and the absence of interface defects ensures the formation of a clean (semi)metal-semiconductor interface with a Schottky barrier, promising for constructing reproducible and reliable optoelectronic devices.

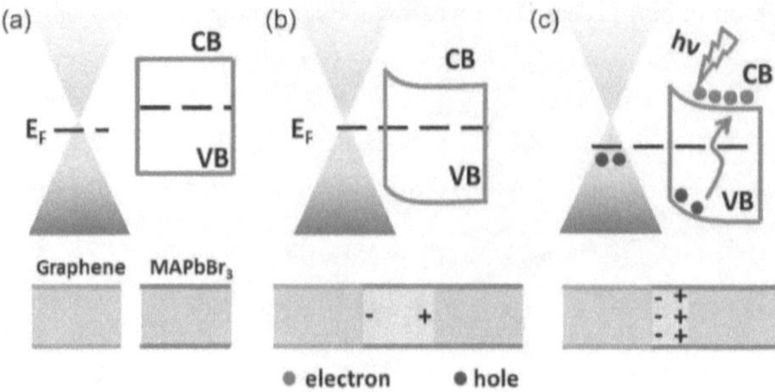

Figure 2.11 Energy band diagrams of graphene and MAPbBr₃ platelets before contact, under dark and light conditions, respectively. Between graphene and MAPbBr₃, the depletion region is highlighted by the yellow color.

Figure 2.11a-c illustrates the band alignment of the MAPbBr₃/graphene under various measurement conditions. According to Raman the spectra, the pristine graphene is slightly p-doped, which means its work function is a little larger than 4.5 eV (Figure 5c). When contact, electrons in MAPbBr₃ flow to graphene to align their Fermi levels and form a depleted space charge area. Thus, the bands of MAPbBr₃ bend upward in the dark condition (Figure 5d). The width of the space charge region (x_w) can be calculated by

Equation 2.3
$$x_w = \sqrt{2\varepsilon_s V_{bi}/qN_d},$$

where V_{bi} is the built-in voltage at the Schottky contact, ε_s is the static permittivity of MAPbBr₃, q is the electron charge, and N_d is doping concentration.[49] From previous results, the carrier concentrations for MAPbBr₃ single crystal under dark and light conditions are $< 5 \times 10^{12}$ cm⁻³ and $\sim 5 \times 10^{16}$ cm⁻³, respectively.[100] Given $\varepsilon_s = 7.5 \times 8.854 \times 10^{-12} F \cdot m^{-1}$,[101] $V_{bi} = (4.5 - 4.0)\ eV = 0.5\ eV$ and $q = 1.6 \times 10^{-19}C$, the x_w of the MAPbBr₃/graphene junction was calculated to be around 6 μm, which is much larger than the thicknesses of the platelets. Thus, MAPbBr$_3$ platelets with thickness less than x_w are completely depleted.

Under the light condition, a number of electrons and holes are generated in the MAPbBr$_3$ platelets, as shown in **Figure 2.11**c. The upward bent valance band can facilitate the hole transfer from MAPbBr$_3$ platelets to graphene, leaving the electrons trapped in MAPbBr$_3$ and leading to more negatively charged surfaces, as shown in the KPFM. Therefore, I_{ds} increases compared to the dark condition and the Dirac voltage shifts to a larger voltage. Moreover, the p-doping concentration (N_d) can be calculated according to:

Equation 2.4
$$N_d = \frac{I_{ph}}{I_{dark}} \frac{\varepsilon_{ox}}{q d_{ox}} \left(V_{Dirac} - V_g\right),$$

where I_{ph}, I_{dark}, ε_{ox} and d_{ox} are photocurrent, dark current, the dielectric permittivity and thickness of the insulating SiO$_2$ layer, respectively. The equation for p-doping concentration is deduced according to a previous report.[102] The details are shown below:

The accumulated charge Q under a gate voltage (V_g) can be calculated as:

Equation 2.5
$$Q \equiv qpA = CV_g ,$$

where p is the hole concentration, A is the surface area of the channel, and C is the capacitance. The capacitance C is given by $C = \varepsilon_{ox} A / d_{ox}$. At a given V_g, the hole concentration without light is

Equation 2.6
$$p_{dark} = \frac{\varepsilon_{ox}}{q d_{ox}} \left(V_{Dirac} - V_g\right).$$

The current in the channel is calculated below according to Ohm's law:

Equation 2.7
$$I_{ds} = qp\mu V_{sd} \frac{W}{L},$$

where q is the electron charge, μ is the carrier mobility, and W and L are the width and length of the channel. V_{sd} is the source-drain voltage. Thus, the p-doping concentration (N_d) from dark to light is derived as:

Equation 2.8
$$N_d = p_{light} - p_{dark} \approx p_{light} = \frac{I_{light}}{I_{dark}} \frac{\varepsilon_{ox}}{q d_{ox}} (V_{Dirac} - V_g).$$

The carrier mobility is assumed constant with and without light illumination.

Based on the transfer data shown in **Figure 2.10**a, an $N_d = 7.1 \times 10^{12}$ cm^{-2} can be calculated under zero V_g, which is almost the same as that derived from the Raman spectra. The increased doping density caused by the photo-generated carriers suppresses x_w by around 100 times (around 60 nm calculated used the same equation in the dark). As a result, the surface of an 83-nm-thick platelet is very different from the thinner platelets.

2.5 Conclusion

In summary, mixed-dimensional MAPbBr$_3$/graphene vdW heterostructures are realized by growing single-crystalline MAPbBr$_3$ platelets on graphene using the one-step CVD method. It was observed that MAPbBr$_3$ platelets selectively grew on graphene surfaces, especially near the edge regions. The strong perovskite/graphene interaction was evidenced by PL quenching and Raman spectroscopy studies, and a p-type doping of graphene with a density of 7.5×10^{12} cm^{-2} was estimated, which was further confirmed by the KPFM and CAFM results. Phototransistors fabricated on the heterostructures revealed an increase of the Dirac voltage, in line with the electron transfer from graphene to perovskite under light illumination. This work provides a new platform of investigating the interfacial coupling in vdW heterojunctions and stimulates further research on their unique properties and potential device applications

Chapter 3 Metal Halide Perovskite and Phosphorus Doped g-C$_3$N$_4$ Bulk Heterojunctions for Air-Stable Photodetectors

3.1 Summary

The charge transfer observed in MAPbBr$_3$ and graphene should be a general phenomenon with the same band structure. It can be utilized to improve the performance of devices like photodetectors. But the graphene is not easy to prepare and its own properties like band structure is not very facile to modify. Therefore, we fabricate photodetectors made of 2D phosphorus-doped graphitic carbon nitride and MLHP bulk heterojunctions. Upon the exfoliation of the synthesized bulk phosphorus-doped graphitic carbon nitride (PCN-B) to nanosheets (PCN-S), the band gap is reduced, and the band structure is demonstrated to match with the perovskite, facilitating the CT process at the junction interface. Owing to the light-alterable built-in barrier at the formed bulk heterojunctions, simple and robust mixing of PCN-S and MLHP presents a convenient and efficient approach to control the photoinduced-interfacial carrier transfer of the junction. As a result, the photocurrent is enhanced due to the fast charge injection and charge separation, while the dark current is significantly suppressed because of the larger barrier. Consequently, the on/off ratio increases to 10^5, and the photodetectivity reaches a value of 10^{13} *Jones*. In addition, the moisture stability of the hybrid film is improved due to the promoted surface hydrophobicity. These results suggest that low-temperature processed PCN-S/MLHP hybrids are promising to fabricate low-cost, reliable, and high-performance MLHP-based photodetectors.

3.2 Introduction and background

Metal halide perovskites have garnered a lot of research interest because of their impressive optical and transport properties such as tunable band gap,[31, 39] high absorption coefficient,[40, 48] long charge carrier lifetime,[27, 32] small exciton binding energies, and high ambipolar mobility.[32, 59] Together with low-temperature preparation capability, these pivotal features make perovskites an excellent candidate for a variety of potential applications, including solar cells,[13, 24] light-emitting diodes,[57, 103] photodetectors,[26, 67] lasers,[31, 37] waveguides,[29, 104] and X-ray imaging.[105, 106] Among these, light detection using perovskites stands out as a promising application and many efforts have been dedicated to it. However, the trade-off between responsivity and detectivity is crucial in this class of materials. While large responsivity requires excellent crystals to obtain a high photocurrent, good detectivity favors defects and grain boundaries to reduce the dark current.[26, 67, 107] Pioneering works confirmed that building a light-alterable energy barrier to control carrier transport with the assistance of foreign materials is one of the possible approaches to resolve this paradox. For example, a heterojunction formed between perovskite and WS_2 layers can increase or decrease the Schottky barrier between electrodes and perovskites in the dark and under light illumination, respectively.[26] Besides, a unique geometry consisting of thousands of serial hopping barriers in a perovskite wire containing insulating organic blocks can remarkably suppress the dark current and conduct carriers through the edges.[67] However, the complex fabrication processes and poor moisture stability greatly restrict their further applications. Therefore, low-cost fabrication of perovskite photodetectors with good stability in air is essential.

Recently, graphitic carbon nitride (g-C_3N_4), as a 2D organic semiconductor with the Fermi level approximately in the middle of the band gap, has emerged as an attractive photocatalyst thanks to its striking features like abundance, stability, and nontoxicity.[108-111] Compared to pristine g-C_3N_4, heteroatom doped g-C_3N_4 has the advantages of tunable band gap and the ability to engineer the electronic structure, leading to a wider light absorption range.[108, 111] This also underlines the feasibility of forming various junctions with other semiconductors. In addition, pristine and doped g-C_3N_4 possess advantages of easy fabrication, metal-free composition, and low cost compared to other 2D materials like MoS_2 and WS_2. Furthermore, a Schottky junction between perovskites and chemically doped g-C_3N_4 may be a viable approach to suppress the dark current and facilitate the interfacial charge transfer (CT) process.

3.3   Experimental section

3.3.1   Material preparation

Preparation of CN-S/PCN-S dispersion: 1 g melamine (Sigma-Aldrich) was dissolved in 100 ml of hot water and 1 g hydroxyethylidene diphosphonic acid (HEDP, 60 wt% in water) is diluted using 5 ml of water to keep the molar ratio of melamine to HEDP around 10:1. Then HEDP solution was added into the melamine solution drop by drop and dried at 105 °C to get the residue. By means of facile thermal polymerization, the residue (heated to 550 °C for 2 h at a heating step of 4 °C/min) was converted into PCN-B. The carrier gas was argon with a speed of 40 sccm. As synthesized PCN-B was then mechanically exfoliated into nanosheets in N,N-Dimethylformamide (DMF) using a probe sonicator with a 10 s ON pulse and a 5 s OFF pulse at an amplitude of 40% in an ice bath for 2 h. The upper clean liquid was collected to get the PCN-S dispersion. Same procedures were followed to get the CN-S dispersion by using only melamine.

Preparation of MAPbI$_{3-x}$Cl$_x$ (MLHP): 0.419 g MAI (Deysol) and 0.245 g lead chloride (Sigma-Aldrich) were dissolved in 1 ml DMF solvent. The mixture was heated at 65 °C overnight under stirring.

Preparation of hybrids: 0.419 g MAI (Deysol) and 0.245 g lead chloride (Sigma-Aldrich) were dissolved in 1 ml PCN-S dispersion and heated at 65 °C overnight under stirring. Same process was followed to prepare CN-S/MLHP hybrid.

3.3.2   Photodetector fabrication

To fabricate photodetectors, the hybrid was spin-coated onto a gold electrode pre-patterned glass substrate. The channel between the two electrodes was 40 μm in length and 600 μm in width. Then the substrate was annealed on a hot plate under 100 °C for 1 h. All these processes were conducted in a glove box filled with N$_2$.

3.3.3   Characterizations

The thermal process to get CN-B or PCN-B was carried in a furnace (Thermal Scientific, Lindberg-Blue-M). A horn probe tip sonicator was used to prepare the dispersions in DMF. XRD patterns were measured on a Bruker D8 Advance diffractometer with Cu Kα (λ = 1.5406 Å). XPS and XPS valance band spectrum measurements were performed using a Kratos Axis Ultra DLD spectrometer with Al Kα radiation ($h\nu$ = 1,486.6 eV). The UV-Vis diffuse reflectance/absorption spectra were recorded by an ultraviolet-vis-NIR spectrophotometer (Cary 5000, Variam Inc.). Mott-Schottky Plots were tested for drop-casted CN-S and PCN-S films on ITO substrates using a CHI660D Electrochemical Analyzer under a frequency of 1221 Hz. Steady-state PL measurements were performed at an excitation wavelength of 532 nm (LabRAM, Horiba-Jobin-Yvon). Time-resolved PL measurements were performed using a pulsed laser diode (405 nm, Horiba-Jobin-Yvon, model (DD-405L, IRF ~65 ps)). The repetition rate of the pulsed laser was fixed at 1 MHz. The used laser energy per pulse was minimized as low as possible to solely excite the perovskite.1 FTIR spectra were measured on Nicolet iS10 (Thermo Scientific). Contact angle tests were conducted on Kruss. The morphologies of the powders or films were examined by a Quantum 600 (FEI Co) SEM and the EDS of the film were obtained using an EDS spectrometer by SEM with an acceleration voltage of 15 kV. The thickness of the sheets and the surfaces of the films were also measured or scanned using AFM (Bruker Dimension ICON). PESA measurement was performed by Riken AC-2 photoelectron spectrometer. I-V curves were measured using a Keithley 4200 semiconductor characterization system connected with a Lakeshore probe station. The light sources were provided by LEDNW LEDs form Metrohm Autolab and the intensity was controlled by the current and external filters. The transient responses were recorded using a digital oscilloscope (Tektronix DSOX3104A). The noise current of photodetectors was measured by a lock-in amplifier (SR830).

3.4　Results and discussions

3.4.1　Synthesis of phosphorus-doped g-C_3N_4 dispersion

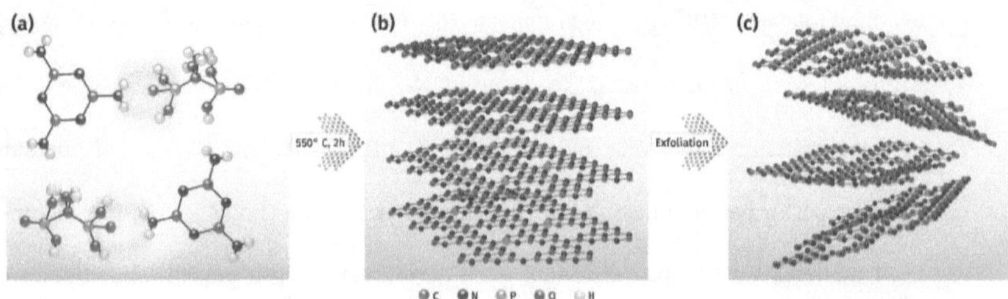

Figure 3.1 Schematic illustration of the PCN-S synthesis process: (a) Structure of the melamine:hydroxyethylidene diphosphonic acid precursors. (b) Transformation of precursors into PCN-B through a thermal polymerization process. (c) Exfoliation of PCN-B into PCN-S in DMF. (The color code is: C, Gray; N, Blue; P, Orange; O, Red; H, White.)

In a typical experiment, bulk $g\text{-}C_3N_4$ (CN-B) and PCN-B were synthesized *via* a bottom-up method; *i.e.* facile thermal polymerization process, using low-cost precursors (melamine for CN-B and extra hydroxyethylidene diphosphonic acid for PCN-B) as shown in **Figure 3.1**. In our approach, hydroxyethylidene diphosphonic acid acts as the source of phosphorus. The pre-mixing of melamine and hydroxyethylidene diphosphonic acid can enhance the interactions between alkaline -NH_2 groups and acidic P-OH groups, which benefit the doping of phosphorus in the $g\text{-}C_3N_4$ backbone during the thermal polymerization process. Such an approach does not require delicate conditions and expensive precursors, which are usually needed in the synthesis of other 2D materials. In addition, a large amount of products in one pot is achievable, which benefits the large-scale applications in devices.

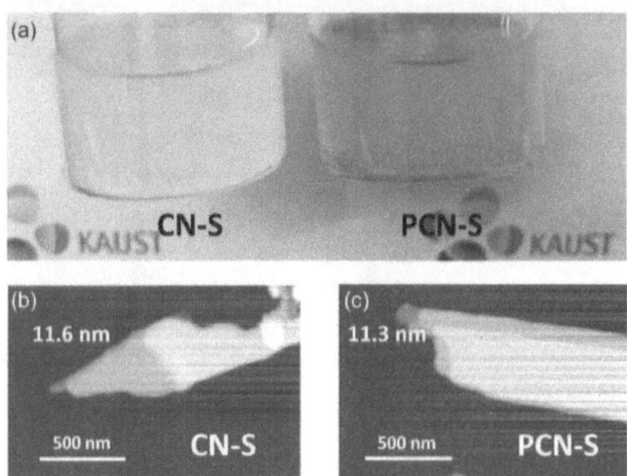

Figure 3.2 (a) Digital image of CN-S and PCN-S dispersions. (b), (c) AFM images of CN-S and PCN-S, respectively.

The synthesized powders were mechanically exfoliated into thin sheets in DMF solvents with the help of probe-sonication. The synthesized CN-S and PCN-S sheets dispersed in DMF solvents are shown in **Figure 3.2**a. Compared to the suspension of CN-S in DMF, the color change of PCN-S dispersion as seen from the digital photographs indicates the successful phosphorus doping in PCN-S. Individual nanosheets can be detected by AFM, as presented in **Figure 3.2**b-c, where both CN-S and PCN-S have a similar thickness of ~11 nm.

3.4.2   Characterization of PCN-S

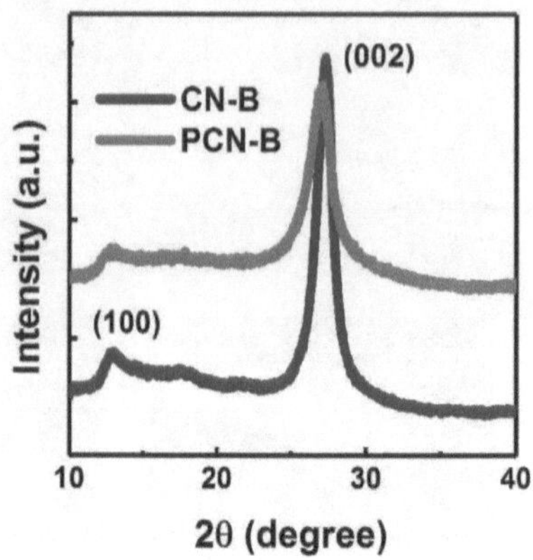

Figure 3.3 XRD patterns of synthesized CN-B and PCN-B powders.

The crystal structure of the as-synthesized PCN-B sample was characterized by XRD as shown in **Figure 3.3**. Two peaks locate at around 13° and 27° can be identified, which is consistent with previous reports.[108] The peaks are attributed to (100) and (002) planes, respectively. A small shift to low-angle of peaks in PCN-B samples compared to CN-B samples indicates a larger interlayer distance due to the large size of phosphorus atoms. Overall, the patterns in PCN-B are similar to those in CN-B, indicating that the crystal structure of PCN-B is largely maintained after the phosphorus doping. Fourier transform infrared (FTIR) spectroscopy spectra of the undoped and doped CN-B are shown in **Figure 3.4**. Exemplary stretches of the C-N heterocycle and breathing modes of the heptazine units are observed in the 1200- 1600 cm-1 region and at 800 cm-1, respectively, for both CN-B and PCN-B. Such results suggest that the pristine conjugated backbone remains unchanged after doping, which is essential for carrier transport. The absence of the vibrational modes of phosphorus-related bonds could be attributed to the low percentage of phosphorus content in the PCN-B.

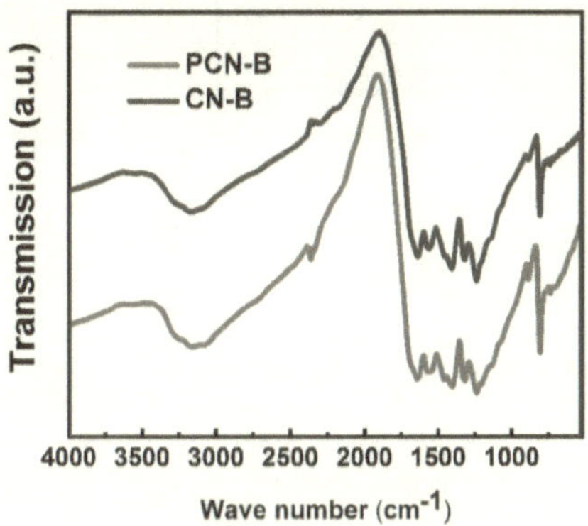

Figure 3.4 FTIR spectra of synthesized CN-B and PCN-B powders.

X-ray photoelectron spectroscopy (XPS) measurement was further utilized to investigate the element details, precise phosphorus content and its preferred doping sites in PCN-B in the synthesized powders as shown in **Figure 3.5**. Only C and N elements were found in the CN-B samples and an extra P element was found in the PCN-B sample. The conjugated heterocycle backbone is again revealed by the high-resolution spectra of the C 1s and N 1s regions as shown in **Figure 3.5**b-c. Noticeably, as displayed in **Figure 3.5**d, a P 2p peak is detected at a binding energy of 133.1 eV for PCN-B, which is consistent with P-N coordination. This indicates that P likely replaces the energy favorable C1 site, as denoted in the inset of **Figure 3.5**d, to form P-N bonds.[108] Furthermore, the precise phosphorus content of around 0.3 wt% was also calculated in PCN-B.

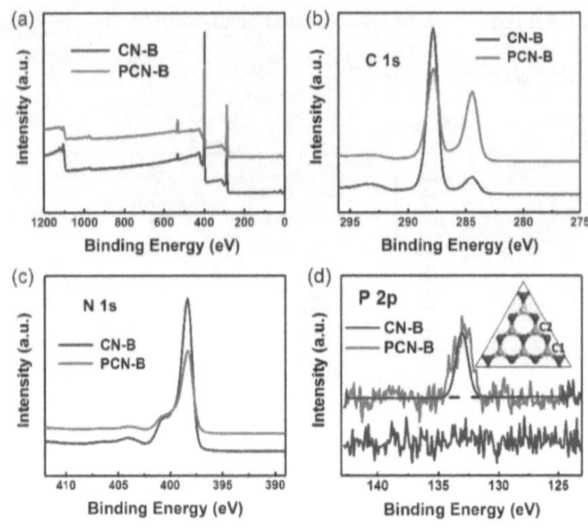

Figure 3.5 XPS spectra of CN-B and PCN-B powders. High-resolution XPS spectra of (b) C 1s, (c) N 1s and (d) P 2p from CN-B and PCN-B powders, respectively. The inset in (d) shows the atomic configuration of phosphorus-doped g-C₃N₄ with two different types of carbon atoms, namely C1 and C2.

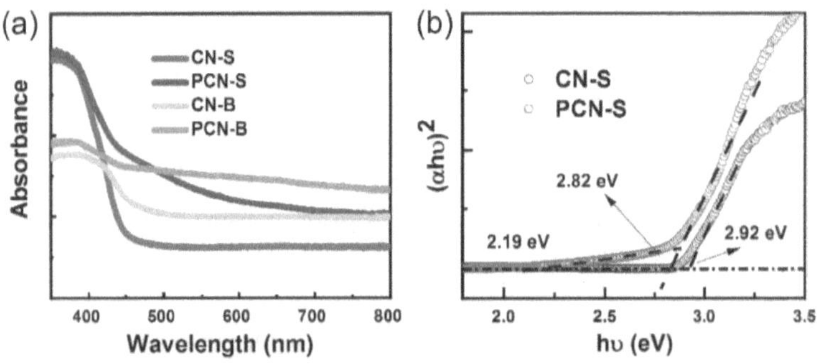

Figure 3.6 (a) UV-vis absorption spectra of CN-B, CN-S, PCN-B and PCN-S. (b) Tauc Plots to deduce the band gaps of CN-S and PCN-S.

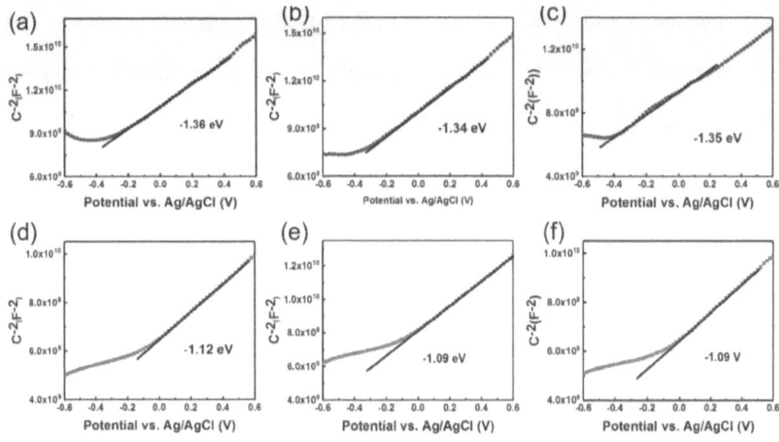

Figure 3.7 Mott-Schottky plots of (a-c) CN-S and (d-f) PCN-S to calculate the average CB.

The band gaps of the CN-S, PCN-S, CN-B and PCN-B were obtained using UV-Vis diffuse reflectance measurement as shown in **Figure 3.6**a. It is clear that the sheets have smaller absorption cut-off values compared to their bulk samples, indicating the successful exfoliations. After exfoliation, the band gaps of either CN-B or PCN-B are narrower than those of CN-S or PCN-S, respectively, which is consistent with previous studies.[108, 112] A red-shifted absorption edge can be observed for PCN-S, suggesting a reduced band gap compared to the CN-S. Most importantly, in contrast to CN-S, the PCN-S demonstrates a clear Urbach absorption tail between 450 nm and 800 nm. This is attributed to the midgap states stemming from the hybridization of C 2s2p, N 2s2p and P 3s3p,[108] illustrating the improved light absorption abilities and modified electronic properties. The values of band gaps are estimated from the Tauc plots in **Figure 3.6**b, using the Kubelka-Munk method.[113] Upon doping, the 2.92 eV band gap of the CN-S is narrowed down to 2.82 eV for PCN-S. The PCN-S, in particular, is found to possess a 2.19 eV transition energy from the valence band (VB) to midgap states, which is on par with previous reports.[108, 111] These midgap states are of great significance acting as hopping centers to afford extra charge carrier transfer at the interface.

The conduction band (CB) positions of CN-S and PCN-S were obtained by using the electrochemical Mott-Schottky plots. Three samples were measured for both CN-S and PCN-S. The results are shown from the capacitance (C)-potential relationship in **Figure 3.7**. Here C^{-2} versus potential *vs.* Ag/AgCl at PH = 6.6 are plotted for better data analysis. The flat band potential can be first determined from the intercept on C^{-2}-axis. Approximate values are collected for both CN-S and PCN-S as we can see. The average flat band potential values are -1.35 eV for CN-S and -1.09 eV for PCN-S *vs.* Ag/AgCl at pH = 6.6, which is -0.73 eV and -0.49 eV *vs.* Ag/AgCl at pH = 0. Thus, the CB of CN-S and PCN-S related to vacuum level can be calculated to be -3.77 eV and -4.02 eV using the reported relationship,[113] respectively.

3.4.3 Preparation of films with hybrids

Various temperatures were tested to anneal the film after spin coating. The PCN-S/MLHP hybrid was used for the test. After spin coating of the hybrid, the films were annealed at 70°, 85°, 100°, and 115°. The resulted films were characterized by XRD as shown **Figure 3.8**. We can find that the film annealed at 100° has the sharpest peaks from the XRD spectra. The crystallization of MLHP grains is not enough below 100° while the grains may start to decompose at 115°. Thus, PCN-S/MLHP hybrid films have achieved the highest crystallinity at an annealing temperature of 100 °C. Then, for comparison, all the studied spin-coated films were first annealed at 100 °C.

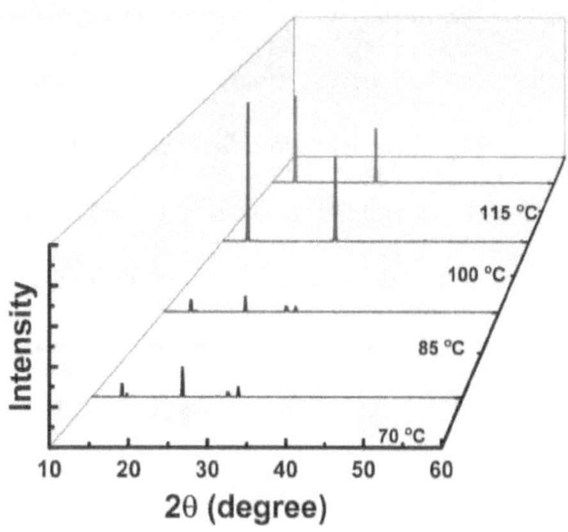

Figure 3.8 XRD spectra of PCN-S/MLHP films annealed at various temperatures.

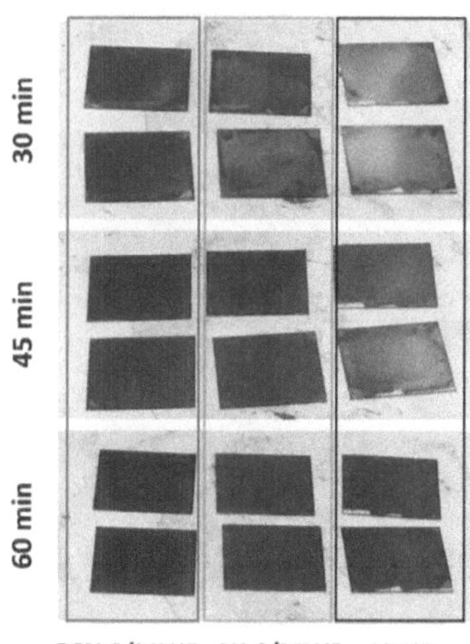

Figure 3.9 Digital photos of MLHP, CN-S/MLHP and PCN-S/MLHP films in annealing at different time. The black color in the PCN-S/MLHP film indicated the faster crystallization to the MLHP phase.

During the experiments, we also observed the crystallization of the films. The speed of the crystallization is reflected from the color change of the film. During the crystallization, the film becomes yellow first and then turns to be dark brown and slight black in a row. The black will fade away to white and develop into black again to be the final MLHP phase. We find that both CN-S and PCN-S can increase the crystallization speed of MLHP, but MLHP crystallizes faster with the presence of PCN-S as illustrated in **Figure 3.9**, in which PCN-S/MLHP turned into black MLHP phase first. Thus, the addition of PCN-S may facilitate the nucleation of MLHP on its surface.

3.4.4 Characterizations of the hybrid films

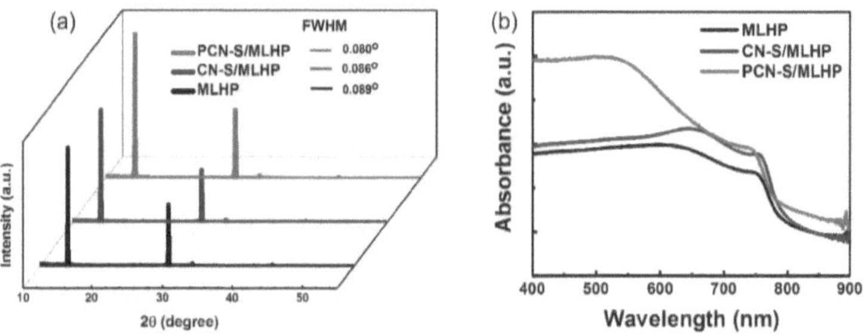

Figure 3.10 (a) Powder XRD patterns of MLHP and hybrids. (b)) UV-vis spectra of MLHP and hybrids.

The corresponding XRD patterns of MLHP and the hybrid films are presented in **Figure 3.10**a. Two main peaks locating at around 14° and 28° are attributed to (110) and (220) planes, respectively. The unchanged patterns demonstrate that the addition of CN-S or PCN-S has no effect on the crystal structure of MLHP. In fact, the full width at half maximum (FWHM) of the (110) peak for the PCN-S/MLHP hybrid film is the smallest, suggesting that the crystallinity of the MLHP is improved. **Figure 3.10**b reveals the UV-vis absorption spectra of the MLHP and the hybrid films. Previous results prove that PCN-S has a higher absorption at around 500 nm due to the extra mid-gap states. As a result, PCN-S/MLHP hybrids have higher absorbance efficiency than CN-S/MLHP and MLHP at shorter wavelength regions (< 600 nm).

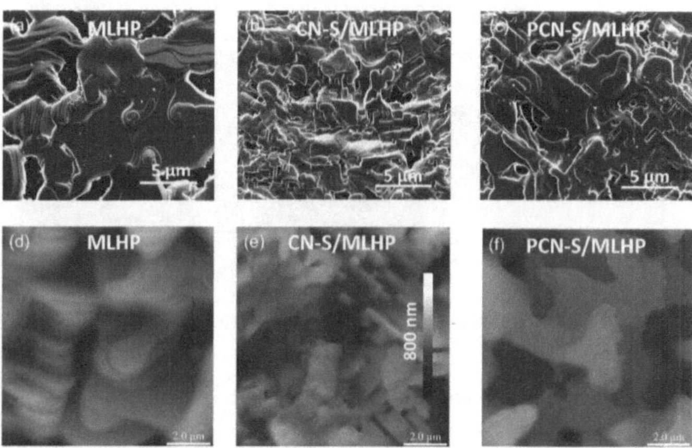

Figure 3.11 (a), (b), (c) SEM images of MLHP, CN-S/MLHP, and PCN-S/MLHP films respectively. (d), (e) (f) AFM images of MLHP, CN-S/MLHP, and PCN-S/MLHP films, respectively.

The SEM images of MLHP and hybrid films in **Figure 3.11**a-c reveal that the irregular MLHP grains become connected, forming micro-cuboid crystals in the hybrid films. The trend is also demonstrated by the AFM images displayed in **Figure 3.11**d-f. As a result, these better-oriented grains further are good to improve the device performance.[114, 115] To further verify the successful mixture of PCN-S in perovskites, the energy dispersive X-ray spectroscopy (EDS) elemental mapping was carried on for PCN-S/MLHP films. The mapping of phosphorus, iodine and lead is shown in **Figure 3.12**. The uniform distribution of these elements validates the uniform blending of MLHP with PCN-S.

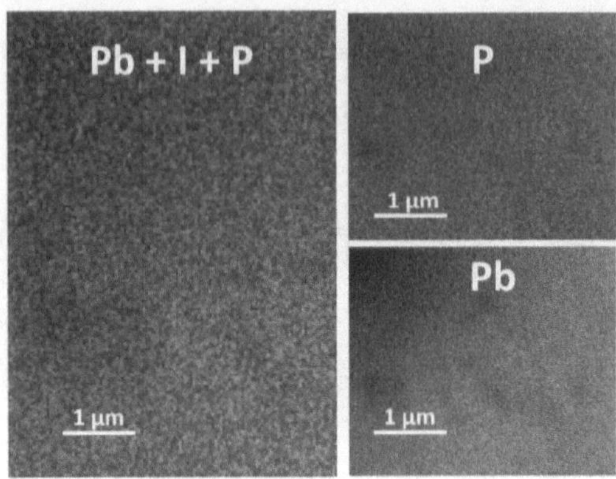

Figure 3.12 EDS mapping of PCN-S/MLHP hybrid films showing lead, iodine and phosphorus elements.

3.4.5 Interactions between MLHP and PCN-S

Using the obtained band gaps and the calculated CB edges, the band structures for CN-S and PCN-S were deduced as constructed in **Figure 3.13**a. The VB positions of CN-S and PCN-S were found to be -6.69 and -6.84 eV, respectively, with additional midgap states in PCN-S located at -4.65 eV. Accordingly, given the well-known MLHP band edges at -3.75 eV (VB) and -5.35 eV (CB),[116] it can be inferred that electrons would transfer from CB of MLHP to CB of CN-S and from CB of MLHP to CB and/or midgap states of PCN-S. This speculation correlates well with the photoluminescence (PL) quenching of the CN-S/MLHP and PCN-S/MLHP hybrids with respect to MLHP as shown in **Figure 3.13**b.

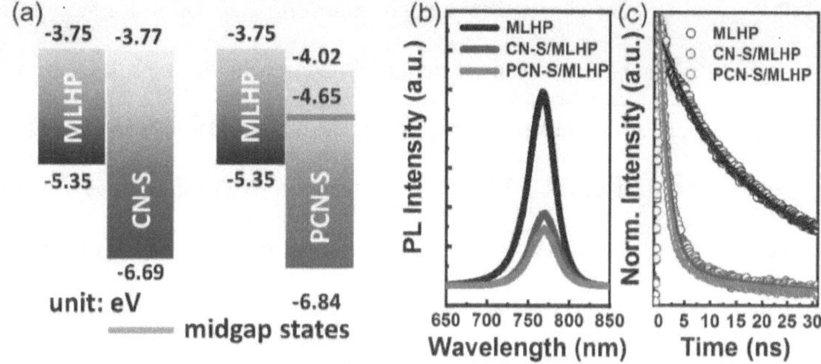

Figure 3.13 (a) Energy band diagram of MLHP combined with CN-S and PCN-S. (b) PL spectra of MLHP, CN-S/MLHP and PCN-S/MLHP hybrids. (c) TRPL spectra of MLHP, CN-S/MLHP and PCN-S/MLHP hybrids. The TRPL intensities are fitted with a bi-exponential decay function over the range of ca. 0 to 200 ns.

This prominent electron transfer taking place in the MLHP hybrids is further confirmed by a clear reduction in the PL lifetimes of the two hybrids *vs.* MLHP as shown in **Figure 3.13**c. To obtain the emission lifetimes of the three studied systems, time-resolved photoluminescence (TRPL) measurements were performed at an excitation wavelength of 405 nm. The curves were fitted by biexponential functions to get the lifetimes (τ_1 and τ_2) and corresponding relative ratio factors (A_1 and A_2). The fitting result is shown in **Table 3-1**, in which τ_{av} is the weighted average lifetime. The TRPL spectrum of the MLHP film shows considerably a long PL lifetime of ca. 26 ns, which has been substantially reduced to few nanoseconds (ca. 2-3 ns) upon the presence of the carbon-based materials. This reduction in the PL lifetime of the perovskite film can be attributed to the charge transfer process from the MLHP to the CN-S/PCN-S materials.[26, 117, 118] In principle, the shortened PL lifetimes in the two hybrids offer great potential for an improved photoresponse. However, the extra midgap states located within the band gap of

PCN-S/MLHP, are expected to support more effortless electron transport than in CN-S/MLHP. This will lead to an efficient exciton dissociation at the interface of PCN-S and MLHP, suggesting a further enhancement in the photodetection behavior of PCN-S/MLHP compared to that of CN-S/MLHP. The enhancement in the photodetection behavior of PCN-S/MLHP is also supported by the enhanced absorption ability of the PCN-S/MLHP hybrid (higher than that of CN-S/MLHP). Hence, based on all the aforementioned, we believe that the cuboid-connected PCN-S/MLHP hybrid film would allow for high-performance photodetectors.[119, 120]

Table 3-1 Fitting parameters of time-resolved photoluminescence spectra of MLHP, CN-S/MLHP and PCN-S/MLHP hybrids. The fitting window is ca. 0 to 200 ns.

Sample	τ_1	A_1	τ_2	A_2	τ_{av}
MLHP	11.74 ± 0.12 ns	50.34%	40.39 ± 0.34 ns	49.66%	**25.97 ns**
CN-S/MLHP	1.99 ± 0.01 ns	92.72%	22.38 ± 0.35 ns	7.28%	**3.47 ns**
PCN-S/MLHP	1.70 ± 0.01 ns	95.33%	25.28 ± 0.31 ns	4.67%	**2.80 ns**

3.4.6 Photodetector performance

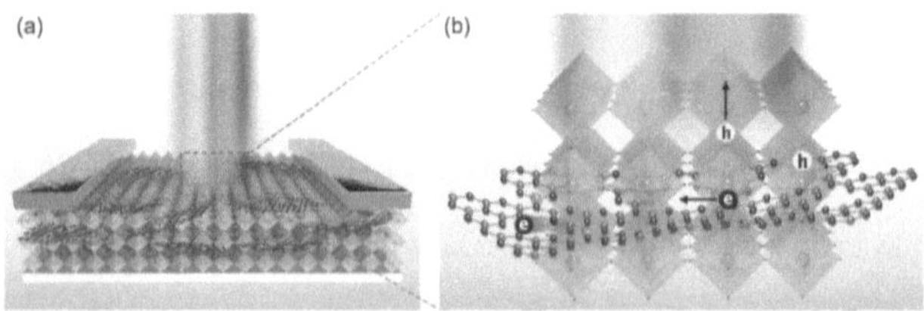

Figure 3.14 (a) Schematic representation of the two-terminal, parallel-type PCN-S/MLHP hybrid thin-film photodetector device under visible light illumination. The drawn PCN-S layer schematic is a simplified representation of the bulk heterojunction. The octahedral stacks represent an MLHP film. (b) The zoomed-in area of the device (dashed red frame in (a)) shows the charge transfer between PCN-S and MLHP in the bulk heterojunctions.

Accordingly, we examined the photodetection performance of MLHP-only as well as its hybrid thin films. The corresponding hybrids were then spin-coated onto glass substrates with pre-patterned gold electrodes to fabricate photodetectors as shown in **Figure 3.14**a. The length and width of the active area in the photodetector are 40 and 600 μm, respectively. The charge transfer depicted in **Figure 3.14**b is beneficial for high photodetector performance.

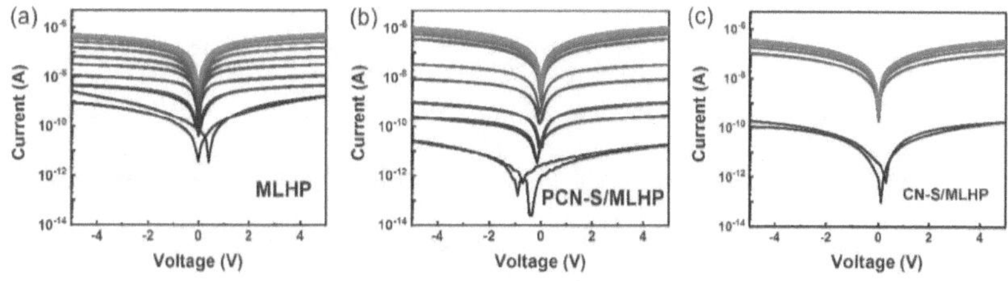

Figure 3.15 Semi-log scale $I–V$ characteristics of (a) MLHP-only, (b) PCN-S/MLHP hybrid, and (c) CN-S/MLHP photodetectors measured in dark and under various white light intensities from 0.5 μW cm^{-2} to 1 mW cm^{-2} with a bias voltage from -5 V to 5 V.

Accordingly, we examined the photodetection performance of MLHP-only as well as its hybrid thin films. The corresponding current-voltage (I-V) curves of devices measured in the dark and under white-light illumination with various power intensities from 0.5 μW cm^{-2} to 1 mW cm^{-2} are shown in **Figure 3.15**. Their performance, including dark current (I_{dark}), current under light illumination (I_{light}), on/off ratio, responsivity (R) and specific detectivity (D^*) is extracted at 5 V under a light intensity of 1 mW cm^{-2} as shown in **Figure 3.16**. R and D^* are calculated from

Equation 3.1

$$R = (I_{light} - \frac{I_{dark)}}{P_{in}}$$

and

Equation 3.2 $$D^* = R/\sqrt{2qI_{dark}/A},$$

where P_{in} is the incident light power, q is the electron charge and A is the effective area of the photodetector.[26] Current noise was measured as shown in **Figure 3.17** to support the utilization of dark current in the calculation of D^* due to the domination of shot noise.[26] Compared to their counterpart MLHP single-layer devices, the photocurrent of the PCN-S/MLHP devices is enhanced several times owing to the midgap states present in the PCN-S, which help in improving the interfacial CT between MLHP and PCN-S. Most importantly, the dark current of the CN-S/MLHP hybrid devices is reduced to 10^{-10} compared to 10^{-9} A for MLHP-only devices, and it continues to be suppressed down to 10^{-11} A in PCN-S/MLHP devices. This could originate from a higher carrier transport barrier at the interface in the dark. Consequently, the on/off ratio of PCN-S/MLHP has reached the highest value of 10^5, which is considered as a significant increase compared to MLHP-only devices (around 10^3).

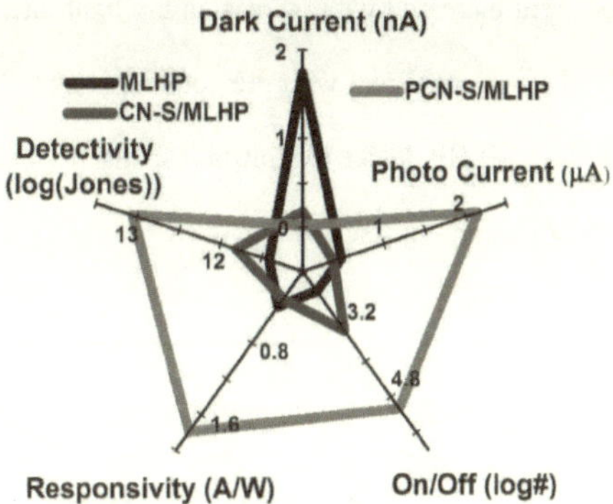

Figure 3.16 Comparison of photocurrent, on/off ratio, responsivity and photodetectivity of devices based on MLHP-only, CN-S/MLHP and PCN-S/MLHP hybrids under a white light intensity of 1 mW cm^{-2} and a bias of 5 V.

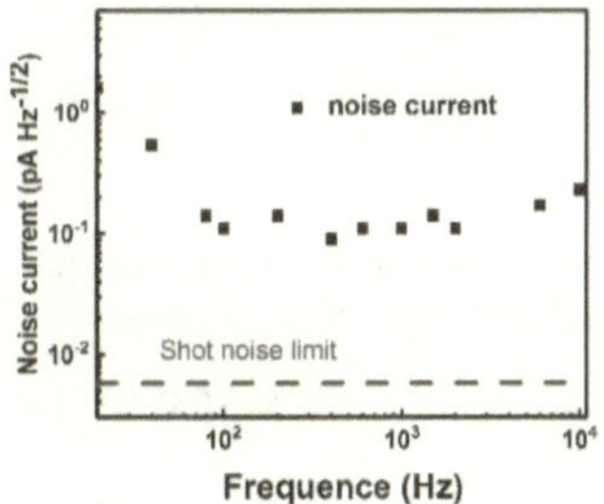

Figure 3.17 Noise current measurement of PCN-S/MLHP based photodetectors.

On the ground of increased photocurrent and suppressed dark current, R and D^* at 5 V under 1 mW cm^{-2} of PCN-S/MLHP hybrid devices are improved to 1.7 A W^{-1} and 1.1 × 10^{13} *Jones*, respectively, as shown in **Figure 3.16**. The two orders of magnitude increased D^* compared to that of MLHP-only devices makes it promising to detect light with ultralow intensity. Further analysis of light intensity-dependent R and D^* variation for a PCN-S/MLHP device measured under a voltage of 5 V is presented in **Figure 3.18**.

The photocurrent increased with increasing the light intensity, while the responsivity has decreased, which is consistent with previous reports.[26, 107] At a light intensity of 0.5 μW cm⁻², the PCN-S/MLHP device exhibits a responsivity of 14 A W⁻¹, which then decreases to 0.86 A W⁻¹ at a light intensity of 5 mW cm⁻².

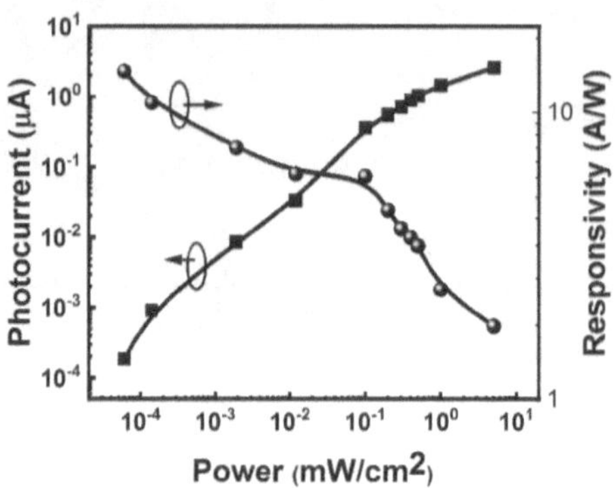

Figure 3.18 White light intensity-dependent photocurrent (left) and photoresponsivity (right) of a PCN-S/MLHP hybrid photodetector under a bias of 5 V.

The photoresponse of PCN-S/MLHP hybrid is demonstrated by wavelength-dependent photocurrent measurements as shown in **Figure 3.19**a. The corresponding R and D^* are presented in **Figure 3.19**b. Linear I-V curves of the PCN-S/MLHP devices exhibit Ohmic-like contacts formed between the electrodes and PCN-S/MLHP hybrids, which is different from the non-linear I-V curves obtained for MLHP-only devices as shown in **Figure 3.19**c. Owing to the small band gap of MLHP, photodetectors made of MLHP and PCN-S/MLHP hybrid exhibit considerably higher R and D^* compared with other reports on other photodetectors measured in the full visible range under a light intensity of 2 mW cm⁻². But the striking feature of these devices is that R is nearly doubled, while D^* is improved by one order of magnitude; this is mostly due to the suppressed dark current and the increased photocurrent aforementioned by adding PCN-S to the MLHP. For instance, at a wavelength of 530 nm, R and D^* are increased from 1.4 A W^{-1} and 0.8 × 10^{12} *Jones* for MLHP to 2.4 A W^{-1} and 7.4 × 10^{12} *Jones* for PCN-S/MLHP, respectively.

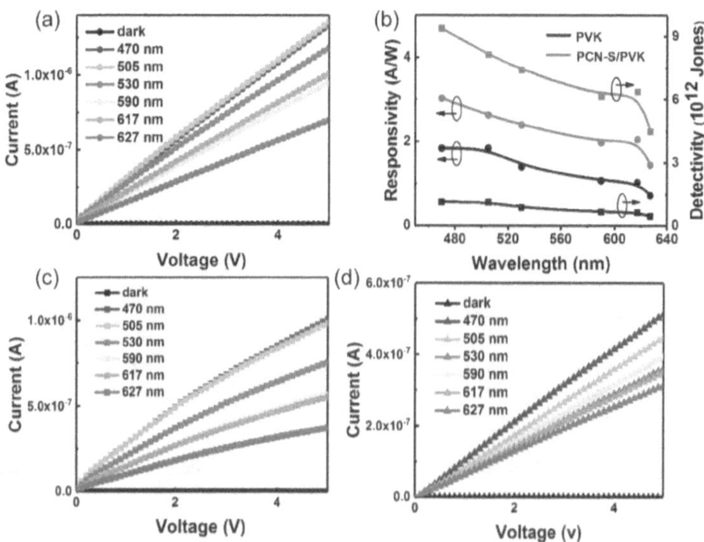

Figure 3.19 (a) Wavelength-dependent I-V curves of PCN-S/MLHP hybrid photodetectors measured under an intensity of 1 mW cm^{-2}. (b) Calculated responsivity and detectivity under a bias of 5 V as a function of wavelength. (c) (d) Wavelength-dependent *I-V* curves of MLHP and CN-S/MLHP based photodetectors under an intensity of 2 mW/cm^2 at a bias of 5 V, respectively.

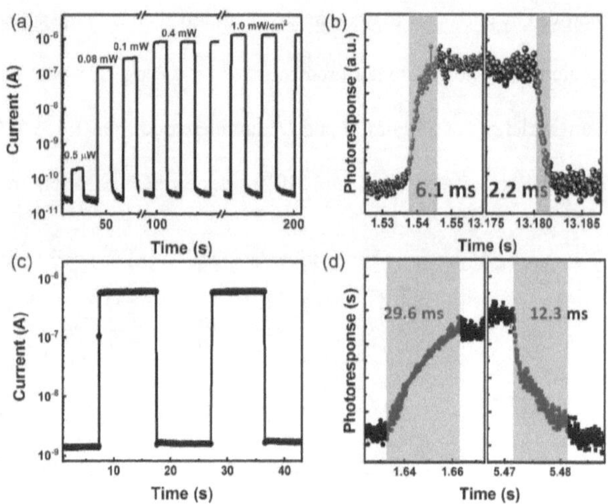

Figure 3.20 Time response. (a) Photoswitching characteristics of PCN-S/MLHP hybrid photodetectors under a bias of 5 V with different intensities of white light under alternating on/off cycles. (b) Time response to give a rise time of 6.1 ms and a decay time of 2.2 ms. (c) (d) Photoswitching behavior and rise/decay times for MLHP photodetectors, respectively.

Reliable and rapid response to light is one of the critical parameters for the performance of photodetectors, which we investigated with pulsed white light illumination. The temporal photoresponse of the PCN-S/MLHP hybrid device is shown in **Figure 3.20**a, which demonstrates reliable photoswitching characteristics under variable light intensities. The hybrid PCN-S/MLHP films promptly respond to white light at the millisecond level as shown in **Figure 3.20**b, resulting in rise and decay times of 6.1 ms and 2.2 ms, respectively. This performance is much faster than the response of MLHP-only devices with a rise and decay times of 29.6 ms and 12.3 ms as shown in **Figure 3.20**c-d, respectively, confirming the fast charge-carrier ejection and separation at the interface between PCN-S and MLHP. A comprehensive comparison of the performance of several planar MLHP-only and other materials based photodetectors is shown in **Table 3-2**. Apparently, our devices exhibit a high on/off ratio, photodetectivity, and adequate responsivity as well as a rapid response time, which lead to net performance advancements in MLHP based photodetectors.

Table 3-2 Performance comparison of previous reports with this work.

Material	Responsivity (A/W)	Detectivity (Jones)	On/off	Response time	Ref.
$MAPbI_3$	0.03	10^{10}	~10	20.5 ms/ 13.8 ms	[121]
$MAPbI_3$	13.6	5.3×10^{12}	$<10^3$	80 µs/ 120 µs	[122]
$WS_2/MAPbI_3$	17	2×10^{12}	10^4	2.7 ms/ 7.5 ms	[26]
$FAPbI_3$	27.6	-	-	12.4 ms/ 17.6 ms	[123]
$MAPbI_3$/Polymer	100	1.5×10^{10}	10^3	-	[76]
$MAPbI_3$	7	-	20	<0.1 s/ <0.1 s	[58]
$MoS_2/MAPbI_3$	3039	1×10^{11}	300	0.45 s/ 0.75 s	[124]
$MAPbI_{3-x}Cl_x$	0.2	-	1550	-	[125]
$MAPbI_3$	20.7	6.5×10^{13}	$>10^4$	17 µs/ 18 µs	[126]
$MAPbI_{3-x}Br_x$	0.055	-	~100	20 µs/ 20 µs	[127]
$MAPbI_3$	4.95	2×10^{13}	10^3	0.1 ms/ 0.1 ms	[128]
$MAPbI_{3-x}Cl_x$	14.5	-	2235	0.2 µs/ 0.7 µs	[129]
$PCN-S/MAPbI_3Cl_{3-x}$	**14**	10^{13}	10^4-10^5	**6.1 ms/ 2.2 ms**	**this work**

3.4.7 Mechanism of improved performance

Our results indicate that the improved performance of the PCN-S/MLHP hybrid photodetectors is linked to the addition of PCN-S to perovskites. To further investigate the influence of PCN-S on both dark and photocurrent, the band alignment of the MLHP and PCN-S heterojunction is plotted for both dark and illuminated situations in **Figure 3.21**a-b, respectively. The band bending directions are determined by the relative Fermi level positions of the two materials. For our perovskites, theoretical calculations and experiments show that MLHP is more *n*-type with a work function of around 4.0 eV.[130] CN-S is reported to be an approximately intrinsic semiconductor, and the Fermi level is further increased in PCN-S due to the phosphorus doping.[111, 131] Photoelectron spectroscopy in air (PESA) and XPS valance band spectra in **Figure 3.22** verify the increased Fermi level (-4.55 eV). From the measured band structure in **Figure 3.13**a, it is evident that there is a great barrier for electron transport from MLHP to PCN-S in the dark, which results in a lower dark current. However, under illumination, free electrons and holes are simultaneously generated inside the MLHP (due to the low exciton binding energies) to reduce the built-in field, and thus the electrons can transfer to PCN-S due to large kinetic energy and density. Consequently, the Fermi level of PCN-S is raised, which minimizes the barrier between the conduction band edges of MLHP and PCN-S. In principle, the barrier could eventually disappear and the midgap states in PCN-S can further facilitate the carrier transfer, which enhances the photocurrent. On the other hand, the high CB, low Fermi level and absence of midgap states in CN-S restrict the efficient carrier transfer between CN-S and MLHP, resulting in lower photocurrent.

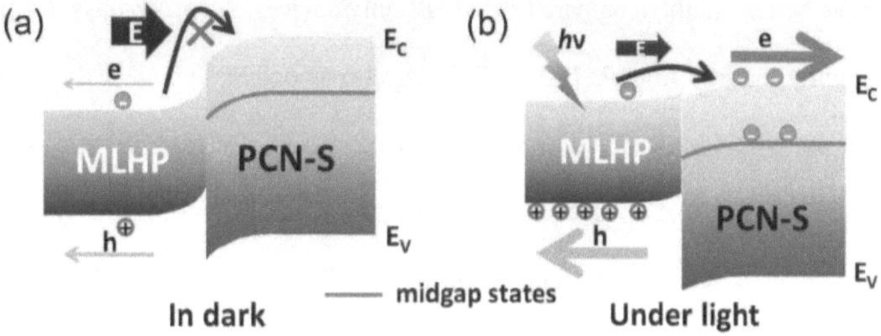

Figure 3.21 Band diagram showing the interface between MLHP and PCN-S (a) in dark and (b) illuminated conditions. The width of the grey or brown arrows indicates the current magnitude and the green line on PCN-S side denotes the position of the midgap states.

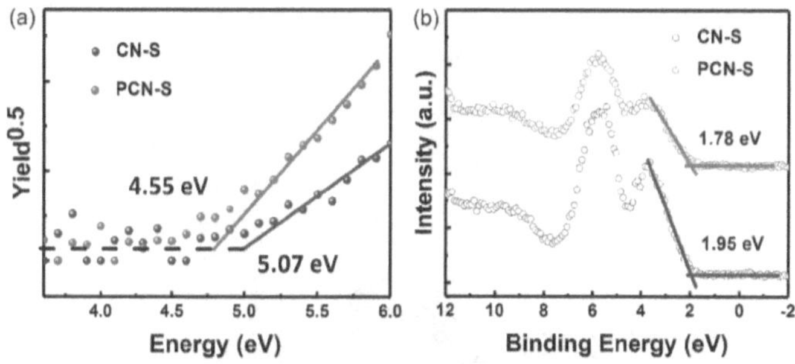

Figure 3.22 (a) PESA measurements and (b) XPS valance band spectra of CN-S and PCN-S.

3.4.8 Improved stability

The stability of metal-halide perovskite photodetectors is a critical issue that must be improved for these materials to realize their full potential. A stability comparison between MLHP and PCN-S/MLHP photodetectors are shown in **Figure 3.23**a-b, where the photocurrent is recorded for a few days, with the samples stored in a petri dish under ambient conditions. It is clear from these data that PCN-S/MLHP hybrid photodetectors have better stability compared to MLHP-only devices. After two days, the photocurrent is slightly reduced from 757 nA to 725 nA, corresponding to a 4.2% drop. In comparison, for the MLHP-only photodetector, the photocurrent decreased from 531 nA to 253 nA, corresponding to 52.3% drop. After seven days, the PCN-S/MLHP hybrid photodetector maintained a relatively high photocurrent of 497 nA, while the MLHP-only device showed almost no photoresponse. This increased stability is evident from the digital image (insets of **Figure 3.23**). Besides, the PCN-S/MLHP hybrid photodetectors exhibit higher stability compared to most solution-processed MLHP photodetectors, whose photocurrents degraded more than 50% in the time frame of 10 minutes to few days.[132-135]

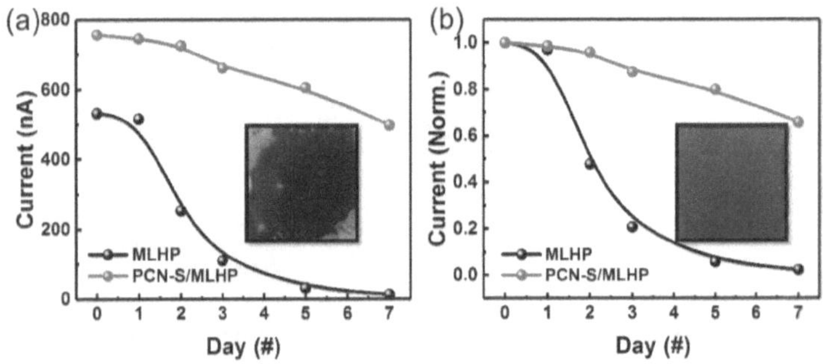

Figure 3.23 (a) Photocurrent under a white light intensity of 0.5 mW cm-1 and (b) normalized photocurrent values of MLHP and PCN-S/MLHP hybrid photodetectors recorded as a function of the number of elapsed days. Insets in (a) and (b) shows, digital images of MLHP and PCN-S/MLHP films, respectively, deposited on glass substrates after one day.

It is generally believed that the poor stability of metal-halide perovskites stems from reactions with water that start at the amorphous grain edges.[134, 136] Thus, a reasonable approach to improve their stability involves modifying the surface wetting behavior (making it more hydrophobic to minimize water absorption on the surface of the MLHP films). Interestingly, the introduction of PCN-S into MLHP seems to achieve this desired result. Contact angle measurements performed on MLHP-only and PCN-S/MLHP hybrid films are shown in **Figure 3.24**a-b, respectively. The measured contact angle of water on MLHP-only films is 22.1°, indicating a very hydrophilic surface. However, this angle becomes significantly smaller due to reaction with water, finally reaching a value of 10.4° after 90 s.[136, 137] In sharp contrast, the PCN-S/MLHP hybrid film has exhibited an initial contact angle of 45.6°, which is significantly larger than that of MLHP, suggesting that the hybrid films are more hydrophobic. Moreover, the contact angle decays much slower compared with MLHP-only films, reaching half of its initial value after 240 s. Therefore, the addition of PCN-S minimizes the moisture diffusion into MLHP grains by modifying the surface wetting behavior and significantly increases the stability of the devices in ambient conditions.

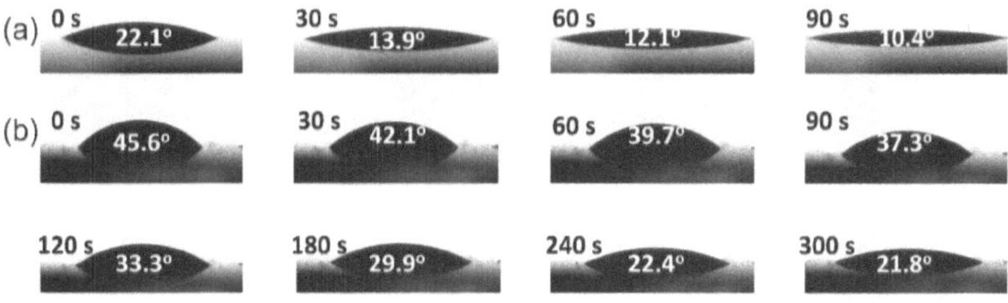

Figure 3.24 The measured contact angle of water on (a) MLHP and (b) PCN-S/MLHP hybrid films as a function of time in seconds.

3.5 Conclusion

In summary, high-performance and air-stable planar photodetectors based on 2D phosphorous-doped g-C$_3$N$_4$ nanosheets (PCN-S) and methylammonium lead tri-halide perovskite (MLHP) bulk heterojunctions have been fabricated using a simple and low-temperature solution processing approach. The hybrid photodetectors demonstrate a high on/off ratio of 10^5, photodetectivity of 10^{13} *Jones* and responsivity of 14 A W^{-1} with reliable and fast photoswitching characteristics. This performance can be ascribed to the suppression of the dark current of MLHP in the presence of PCN-S, and the efficient charge transfer processes at the interfaces between PCN-S and MLHP. Furthermore, the PCN-S/MLHP hybrid photodetectors exhibit increased moisture resistance thanks to their improved surface hydrophobicity. Our results show a new approach to fabricate air-stable efficient perovskite photodetectors by creating bulk heterojunctions with 2D materials. The low-cost fabrication process and high detector performance open a novel window for practical devices.

Chapter 4 Plasmonic MAPbI$_3$/Nb$_2$CT$_x$ Heterostructures for Visible-Near Infrared Photodiodes

4.1 Summary

The previous results show that we can prepare perovskite photodetectors with improved and more stable performance using appropriate heterostructures. However, the detection band is still limited to the visible range. It is desirable to broaden the operation band. Our method is to introduce the surface plasmonic absorption of some MXenes in IR region into the perovskite detector. We fabricate vertical photodiodes with methylammonium lead tri-halide perovskite/MXene (MAPbI$_3$/Nb$_2$CT$_x$) heterostructures to detect visible-NIR lights. The device structure is Au/MAPbI$_3$/Nb$_2$CT$_x$/Au. The MXene used is Nb$_2$CTx because its plasmonic absorption peak is in the NIR region. The MAPbI$_3$ layer acts as not only a barrier layer with a matched band diagram but also a light-absorbing layer in the visible range to broaden the detection range of the photodiode. Therefore, on the one hand, the heterostructure exhibits self-powered performance under white light illumination with a responsivity of 0.25 A/W. On the other hand, the plasmon-induced hot carriers in Nb$_2$CT$_x$ can transfer into perovskite layers, leading to photocurrents under NIR illumination even without external bias. With the perovskite layer, the dark current is significantly suppressed from 10^{-4} A in a control planar Nb$_2$CT$_x$-only device to 10^{-11} A, raising a much enlarged on/off ratio ($<$ 2 to $\sim 10^3$). Most importantly, the response time decreases tremendously (20 s to $<$ 30 ms) with the MAPbI$_3$ layer. The MAPbI$_3$/Nb$_2$CT$_x$ interface plays a crucial role in the charge transfer. The bonding between the surface groups in Nb$_2$CT$_x$ and undercoordinated Pb^{2+} ions in MAPbI$_3$ surface leads to the passivation of their interface, reaching a reduced trap density to benefit efficient and speeded charge transfer.

4.1   Introduction and background

MXenes, as a growing family of transition-metal two-dimensional (2D) materials, have attracted increasing research interest because of their wet processability and appealing physicochemical properties. In general, MXenes have the formula of $M_{n+1}X_nT_x$, where M is an early transition metal, A is C, N, or both, and T represents the surface groups (-OH, -O, -F).[138, 139] In recent years, MXenes have been widely applied in supercapacitors,[140, 141] batteries,[142, 143] electromagnetic interference shielding,[144] and electronics/optoelectronics.[145, 146] Among them, the application of MXenes in optoelectronics stands out promisingly and has been verified by many pioneering works.[145, 147-150] Taking advantage of the excellent conductivity and matched work functions, our colleagues used $Ti_3C_2T_x$ as electrodes for thin-film transistors.[146] Later, flexible transparent electrodes/buffer layers were achieved in solar cells, light-emitting diodes, photodetectors, and phototransistors.[148-150] By then, the applications of MXenes are mainly limited in electrodes and buffer layers. With the exploration of plasmonic effects,[151, 152] the strong light-matter interactions in MXenes under specific frequencies make it possible for MXenes to function as light-absorbing materials in photodetectors. Such observations extensively initiate new roles of MXenes in optoelectronic devices.

MXenes have presented broad plasmonic absorption features from visible to near-infrared (NIR) regions based on their carrier concentrations ($10^{22} - 10^{19}$ cm^{-3}).[151] However, their photodetectors have been achieved only in the visible region, such as Mo_2CT_x.[145]

Thus, more efforts should be exerted in longer wavelengths like the NIR. Besides, for planar photodetectors, the dark current is usually high due to the relatively low resistivity in MXenes, markedly restricting the on/off ratio and photodetectivity. Therefore, new device configurations are demanded to suppress the dark current. A strategy is to fabricate a vertical device and introduce a barrier layer with a matched band diagram.[153, 154] The barrier can be either an insulating layer or a Schottky heterojunction to select hot carriers induced by plasmons under illumination and block off carriers in the dark. With this scheme, successful short-wavelength IR photodetectors based on silicon/aluminum Schottky junctions have been achieved with noticeable performance.[155, 156] Therefore, establishing a barrier for Nb_2CT_x is likely to reduce the photocurrent and improve the performance.

4.2 Experimental section

4.2.1 Material preparation

Synthesis of Nb_2CT_x: 1 g Nb_2AlC powders were added into 20 ml HF aqueous solution (49 wt. %, Sigma-Aldrich) at 55 °C for 48 h with stirring. The resulting suspension was washed by deionized water through several rounds of centrifugation and decantation until a pH of 6-7 was reached. Following the centrifugation/decantation process, the solid sediments were re-dispersed into 20 ml of tetramethylammonium hydroxide (TMAOH, Sigma-Aldrich) aqueous solution (10 wt. %) and left under stirring for 24 h at room temperature for delamination. Afterward, the mixtures underwent another four rounds of centrifugation/decantation. The final sediments were re-dispersed in deionized water and centrifuged at 6000 rpm for 5 min. The supernatant solutions were collected and used for the thin film preparation.

Preparation of Nb_2CT_x thin films: The spray-coating method was used for the preparation of Nb_2CT_x films. Glass and sapphire substrates were first washed by detergents, deionized water, acetone, and isopropanol for 15 min sequentially. Before coating, the substrates were treated by UV-O_3 for 15 min. Nb_2CT_x dispersion with a concentration of 1.5 mg/ml was loaded into the spray gun for the coating. The thickness of the film was controlled by the spray time. The film was dried in a high vacuum overnight and annealed at 100 °C for 5 min before use.

Preparation of $MAPbI_3$ solution: 0.553 g PbI_2 (Sigma-Aldrich) and 0.191 g CH_3NH_3I (Greatcell Energy) were added into N, N-Dimethylformamide/Dimethyl sulfoxide (DMF:DMSO = 4: 1, v:v) mixed solvents. The mixture was stirred and heated at 60 °C for 5 hours. After that, the solution was filtered by 0.22 µm syringe PTFE filters.

Preparation of TiO_2 sol: 370 µL titanium isopropoxide (Sigma-Aldrich) was added into 5 ml anhydrous ethanol. Then 37 µL HCl aqueous solution was dropped into the mixture. The final mixture was stirred at room temperature for 2 h and aged for 24 h.

4.2.2 Device fabrication

Photodetector fabrication: $MAPbI_3$ solution was first spin-coated onto Nb_2CT_x films. At the last 10 s of the spin-coating, 500 µL anti-solvent (chlorobenzene) was dropped onto the substrates. The films (around 800 nm) were annealed at 100 °C for 10 min. Then gold electrodes (60 nm) were deposited with a shadow mask using an e-beam evaporator.

For passivated devices, they were annealed at 100 °C again and aged for at least 24 h in the glove box before measurements. For as-prepared samples, this step was omitted. For planar Nb_2CT_x photodetectors, gold electrodes were deposited on Nb_2CT_x films using an e-beam evaporator with a mask. The device was annealed at 100 °C if needed for comparison.

SCLC device fabrication: For Nb_2CT_x samples, a phenyl-C61-butyric acid methyl ester (PCBM) layer was spin-coated onto MAPbI$_3$ and followed by the electrode deposition and other steps. For TiO_2 samples, TiO_2 sol was first spin-coated onto ITO substrates and annealed at 500 °C for 30 min in air. Other steps were the same as Nb_2CT_x samples.

4.2.3   Characterization

Material Characterization: Surfaces of MAPbI$_3$ film and Nb$_2$CTx films/sheets were recorded by AFM (Bruker Dimension ICON). The thickness of Nb_2CT_x films was measured by AFM after scratching the surface. The thickness of the MAPbI$_3$ was measured by a profilometer (Veeco Dektak 150). The surface morphology of MAPbI$_3$ film was also observed by SEM (Quattro ESEM, Thermo Fisher Scientific). XRD patterns were obtained by a powder X-ray diffractometer (Bruker D8 Advance, Cu Kα radiation). UV-vis-NIR transmittance spectra of Nb_2CT_x films on sapphire substrates were collected using Perkin Elmer Lambda 950. The UPS and XPS spectra of Nb_2CT_x were obtained using Kratos AXIS Ultra DLD. PL spectra and Raman shift were recorded by Witec Alpha300 with the excitation wavelength of 532 nm. The sheet resistances of Nb_2CT_x films were measured by a 4-point probe setup (CMT-SR2000N, Advanced Instrument Technology). Spectroscopic ellipsometry measurements were carried on an ellipsometer with the DeltaPsi2 software (UVISEL, Horiba Jobin Yvon).

Device measurement: All the measurements were carried in ambient conditions. *I-V* and *I-t* (time) curves were measured using a Keithley 4200 semiconductor characterization system connected with a Lakeshore probe station. A white LED was used as the white light source, and a 1064-nm laser (LD-WL206, Changchun New Industries Optoelectronics Tech. Co.) was used as the NIR source. The light intensity was measured by a power meter (PM200, Thorlabs) with detectors (S122C and S120VC, Thorlabs). The frequency/voltage-dependent capacitance and conductance were obtained using an LCR meter (E4980A, Agilent). A direct voltage of 0.5 V coupled with a 50 mV AC signal was applied to get the capacitance-frequency curves.

4.3   Result and discussions

4.3.1   Characterization of Nb_2CT_x layers

Nb_2CT_x is a 2-1 type MXene, which means the stoichiometric numbers of Nb and C are 2 and 1, respectively. The lattice structure of Nb_2CT_x is shown in **Figure 4.1**. It has a close-packed crystal structure, in which Nb atoms follow a hexagonal close-packed stack with an ABABAB ordering and C atoms fill parts of the octahedral interstitial sites. The existence of surface groups (T) makes it convenient to delaminate single layers with the help of intercalating agents from bulk materials. In our case, bulk Nb_2CT_x is first synthesized by etching Nb_2AlC using HF acid, and then we choose TMAOH as the intercalating agent to prepare monolayer Nb_2CT_x sheets.

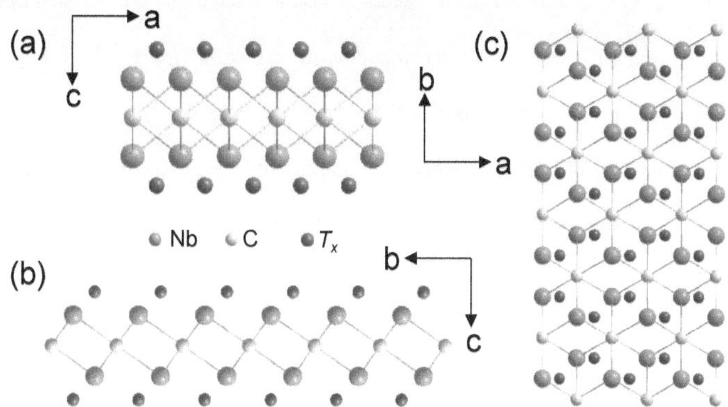

Figure 4.1 Lattice structure of Nb2CTx from different angles. The surface group T can be –O, -OH, and –F.

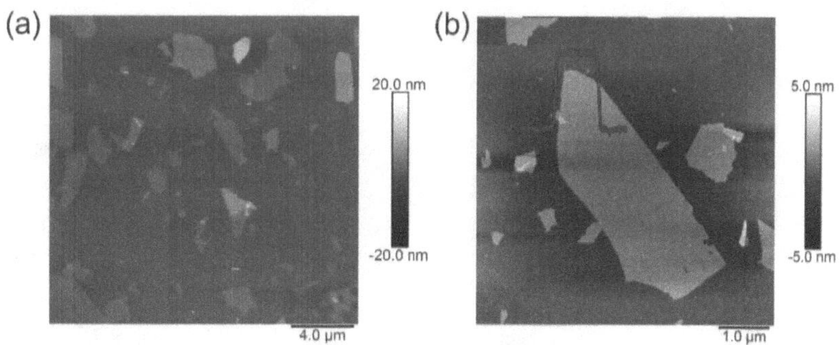

Figure 4.2 AFM images of single-layer Nb_2CT_x sheets in (a) large and (b) small scales. The inset in (b) indicates the thickness of a sinlge-layer Nb_2CT_x is 2.1 nm.

Single Nb_2CT_x sheets observed by AFM are shown in **Figure 4.2**. The lateral size of the sheets can reach 4 µm and the thickness of a single layer is 2.1 nm. In the observed window, almost all sheets are single-layered, indicated the successful delamination process. **Figure 4.3** show the Raman and XPS spectra of Nb_2CT_x sheets. In the Raman spectra, the

E_g peak at around 120 cm^{-1} and A_{1g} peak at around 264 cm^{-1} related to Nb-C bonds can be observed, which is in accordance with previous reports.[157, 158] From the XPS spectra, elements including Nb, C, O, F, N can be discovered, which match the composition of Nb_2CT_x. We can also notice that the peak of N is pretty small, suggesting the tiny amount of N element and the intercalants (TMAOH) were washed away sufficiently.

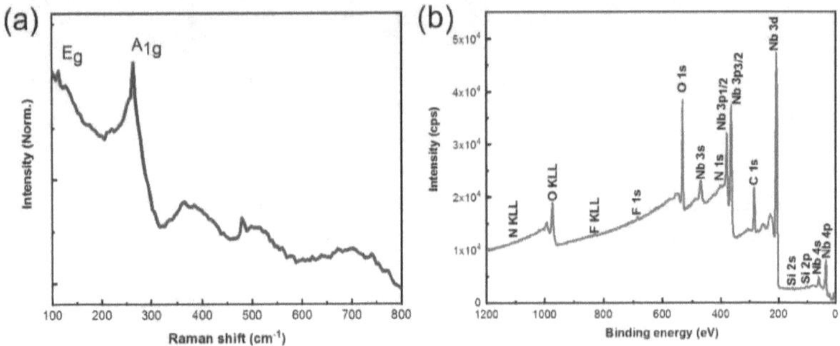

Figure 4.3 Raman spectrum of Nb_2CT_x sheets. (d) Wide-scan XPS spectrum of Nb_2CT_x sheets.

The crystal structure of the sheets were characterized by XRD as shown in **Figure 4.4**. Series peaks from (002) to (0010) can be found, suggesting the pure phase of Nb_2CT_x and the stacked sheets in the film. From the sharp (002) peak at 7.35°, the c lattice constant value is calculated to be 2.4 nm, which is similar to the thickness from AFM.

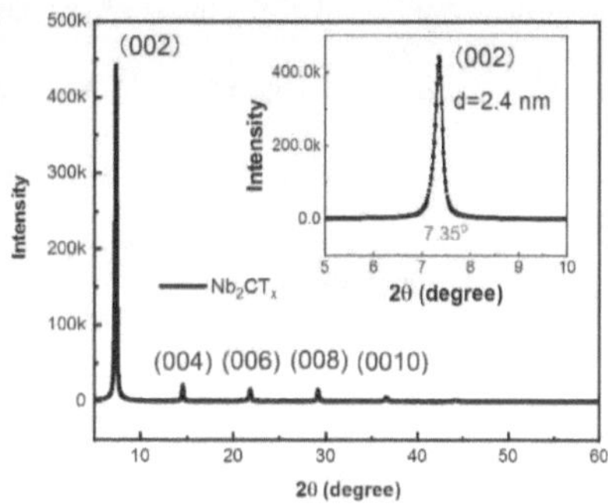

Figure 4.4 XRD pattern of Nb_2CT_x. The inset is the enlarge (002) peak locating at 7.35°.

4.3.2 Preparation of Nb_2CT_x films

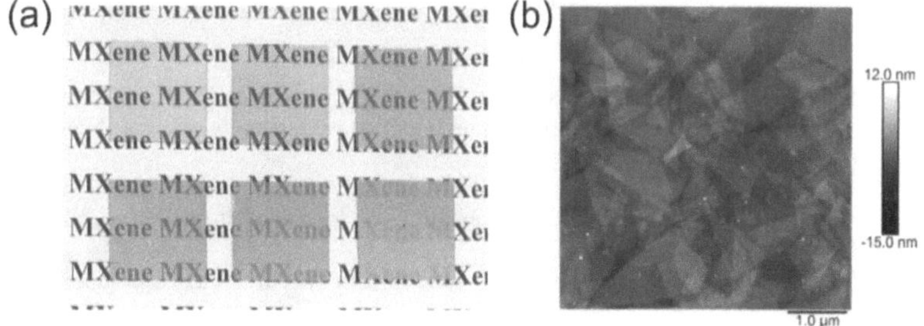

Figure 4.5 Spray-coated Nb_2CT_x films. (a) Digital photo of Nb_2CT_x films with different thicknesses. From left-top to bottom-right, the thicknesses are 8 nm, 12 nm, 18 nm, 22 nm, 37 nm, and 61 nm, respectively. (b) AFM height image of a Nb_2CT_x film surface.

Nb_2CT_x films on glass substrates or sapphire substrates were prepared by spray-coating. The thickness of the film is controlled by the concentration and coating time. **Figure 4.5**a shows the digital photo of prepared Nb_2CT_x films on glass substrates with different thicknesses. The thicknesses of these films were measured by AFM after scratching the surface. A metal-like grey luster can be observed in the thickest film. **Figure 4.5**b is the AFM height image of a Nb_2CT_x film. The root-mean-square (RMS) roughness is less than 1 nm in a 5×5 µm area, indicating a uniform and smooth Nb_2CT_x surface.

4.3.3   Properties of Nb_2CT_x films

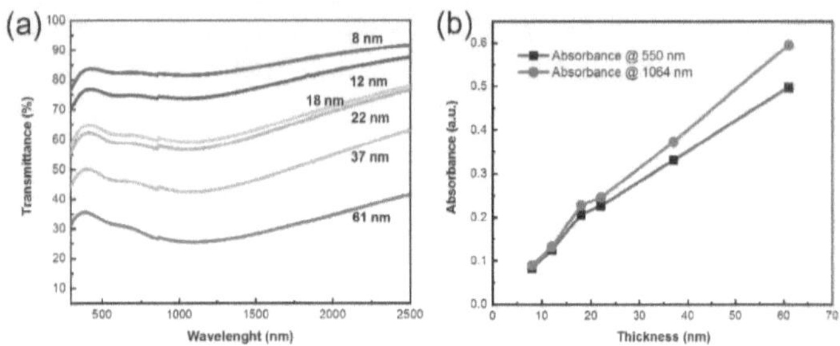

Figure 4.6 Optical properties of Nb_2CT_x. (a) Transmittance spectra of Nb_2CT_x films with different thicknesses. (b) Approximate linear correlation between absorbance at 550/1064 nm and thickness of Nb_2CT_x films.

Figure 4.6a shows their transmittance spectra from visible to NIR region. The transmittance at 550 nm is 81.5% for the thickness of 8 nm and continuously reduces to 25.4% for the thickness of 61 nm. The plots of the absorbance at 550 nm representing the visible and 1064 nm representing the NIR versus thickness shown in **Figure 4.6**b reveal approximately linear correlations as expected from the Beer-Lambert law.[152] In addition, a broad absorption peak centered at 1100 nm can be identified. Such a peak at around 750 nm is usually observed in $Ti_3C_2T_x$ and attributed to the plasmonic absorption.[151, 152, 159] It is reported that this plasmonic peak is strongly related to the carrier concentration and a lower carrier concentration results in a larger peak wavelength. The average carrier concentration in Nb_2CT_x is 3×10^{20} cm^{-3} measured by Hall effects, which is much less than

$Ti_3C_2T_x$ (at the level of 10^{22} cm^{-3}).[151, 152] Therefore, the hypothesis that the plasmonic effect exists in Nb_2CT_x can be proposed and the corresponding absorption peak is red-shifted compared to $Ti_3C_2T_x$ due to the lower carrier concentration.

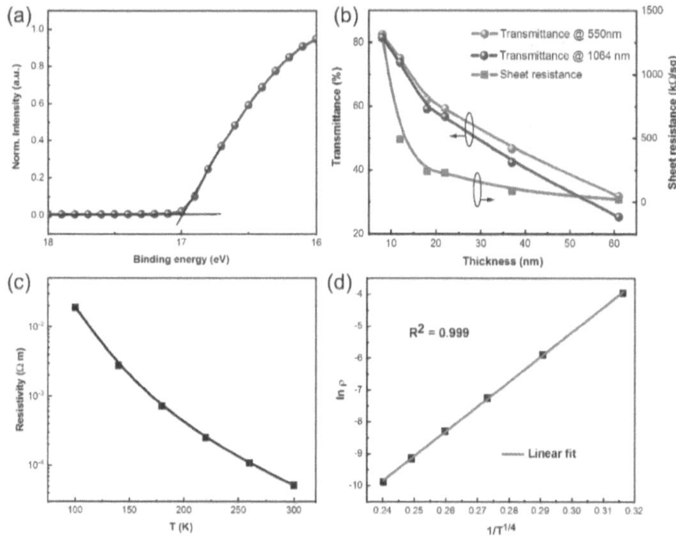

Figure 4.7 Electrical properties of Nb_2CT_x films. (a) UPS measurement to show the work function of Nb_2CT_x. (b) Transmittance at 550/1064 nm and sheet resistances of Nb_2CT_x films with various thicknesses. (c) Temperature-dependent resistivity of a Nb_2CT_x film. (d) Fitting of $\ln\rho$ versus $1/T^{1/4}$.

The work function of Nb_2CT_x is measured by ultraviolet photoelectron spectroscopy (UPS) as shown in **Figure 4.7**a. The high-energy cut-off is 17 eV, giving the work function of Nb_2CT_x to be 4.2 eV. The thickness-dependent sheet resistance and transmittance at 550/1064 nm are shown in **Figure 4.7**b. For a 61-nm-thick Nb_2CT_x film, the sheet resistance is around 21 kΩ/sq with a transmittance of 25.4% at 550 nm. When the thickness reduces to 8 nm, the sheet resistance dramatically increases to 1300 kΩ/sq but with high transmittance. Therefore, a compromise between transmittance and sheet resistance appears for optoelectronic devices. The thickness we used for constructing the $MAPbI_3/Nb_2CT_x$ heterostructure is near 20 nm in this regard. Compared to $Ti_3C_2T_x$, the sheet resistance of Nb_2CT_x is 10^2-10^4 times higher at the same transmittance. The low conductivity of Nb_2CT_x makes the dark current low naturally for optoelectronic devices. The temperature-dependent resistivity in **Figure 4.7**c reveals that the resistivity dramatically increases from 300 K to 100 K. The fitting plot in **Figure 4.7**d shows that the carrier transportation behavior in Nb_2CT_x films likely follows a 3D variable range hopping model,[152] unlike typical metals or highly-doped semiconductors with plasmonic effects.

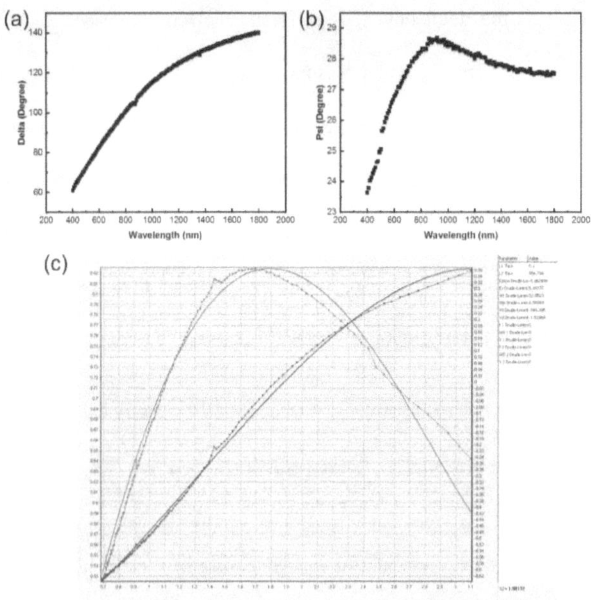

Figure 4.8 Spectroscopy ellipsometry measurements. (a) Delta (Δ), (b) Psi (Ψ) obtained from ellipsometry, and (c) Screenshot of the fitting results using the DeltaPsi2 software.

Spectroscopy ellipsometry measurements were conducted to acquire the optical constants of Nb_2CT_x to check the plasmonic effects. Nb_2CT_x films on n-type silicon wafers were used for the ellipsometry measurements. For that, a structure of $Nb_2CT_x/SiO_x/Si$ was used for the fitting. The theoretical model, utilized to fit the parameter of Delta (Δ) and Psi (Ψ), involves a Drude oscillator and a single Lorentz oscillator, which can be expressed as:

Equation 4.1

$$\varepsilon(\omega) = \varepsilon_\infty + \frac{\omega_p^2}{-\omega^2 + i \cdot \Gamma_d \cdot \omega} + \frac{(\varepsilon_s - \varepsilon_\infty) \cdot \omega_t^2}{\omega_t^2 - \omega^2 + i \cdot \Gamma_0 \cdot \omega}$$

where, ω is the frequency, ε_s is the static dielectric constant, ε_∞ is the high-frequency dielectric constant, ω_p is the plasma frequency, Γ_d is the collision frequency, ω_t is the resonant frequency of the single Lorentz oscillator, Γ_0 is the broadening of the single Lorentz oscillator, and i is the imaginary unit. The fitting results are shown in **Table 4-1**. The value of χ_2, which describe the fitting results (the low the better), is only 1.98, indicating a good fitting. Also, the fitting-estimated thickness of Nb_2CT_x film correlates well with the thickness obtained by AFM (around 90 nm).

Table 4-1 Fitting results of the spectroscopy ellipsometry data

χ_2	Thickness of SiO_x	Thickness of Nb_2CT_x	ε_∞	ε_s	ω_p	ω_t	Γ_d	Γ_0
1.98	0.1 nm	85.7± 11.7 nm	-0.46	5.44	6.55	52.85	-1.52	-349.4

Figure 4.9a presents the optical constants n (refractive index) and k (extinction coefficient) of Nb_2CT_x films from analysis of spectroscopy ellipsometry data in the range of 400- 1800 nm. A Drude and a single Lorentz oscillator are employed for the data interpretation as mentioned above. Increased k values in longer wavelengths can be observed, indicating the absorption in the NIR region. The permittivity ($\varepsilon= \varepsilon_1+ i\varepsilon_2$, where ε_1 is the real part and ε_2 is the imaginary part) is shown in **Figure 4.9**b. Clearly, ε_1 becomes negative from 520 nm to 1590 nm, validating the plasmonic effects in this range and matching the absorption peak in Figure 2b. Being in the same MXene family, $Ti_3C_2T_x$ and Mo_2CT_x have been verified to be plasmonic materials.[145, 152] Therefore, plasmons exist in Nb_2CT_x and can be utilized for photodetection in the NIR region.

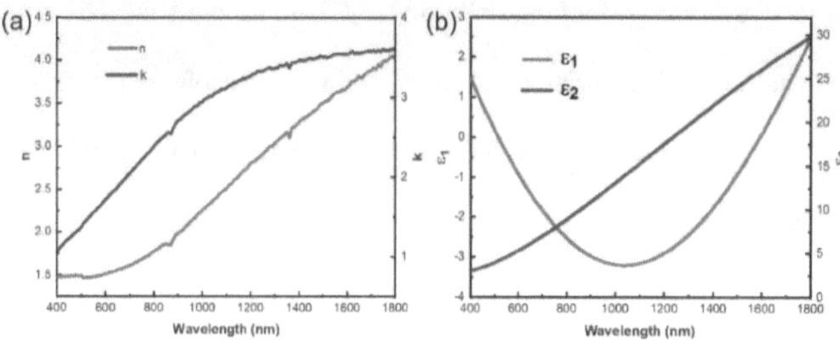

Figure 4.9 (a) Optical constants (n, k) and (b) complex permittivity (ε_1, ε_2) of a Nb_2CT_x film.

4.3.4   Chracterization of $MAPbI_3/Nb_2CT_x$ heterostructures

Figure 4.10a presents the XRD patterns of $MAPbI_3$ and $MAPbI_3/Nb_2CT_x$ heterostructures. After the deposition of $MAPbI_3$, the (002) peak for Nb_2CT_x is slightly shifted to high angles, indicating a deceased c-lattice parameter and a denser film. Even weaker peaks of (006) at 21.6° and (008) at 29.1° of Nb_2CT_x can be observed in the heterostructure. Overall, the patterns for $MAPbI_3/Nb_2CT_x$ are combinations of $MAPbI_3$ and Nb_2CT_x without new peaks and no apparent damage happens for the Nb_2CT_x layer. **Figure 4.10**b-c show the SEM and AFM images of $MAPbI_3$ films on Nb_2CT_x films. Clear $MAPbI_3$ rains can be observed in the SEM image. The average grain size is around 0.4 μm, and the RMS of the $MAPbI_3$ film roughness is 12 nm, which is much less than the film thickness of 800 nm. Such a smooth, dense, and pinhole-free film is suitable for low leakage current devices. Since work function of Nb_2CT_x is 4.2 eV measured by UPS above, the band diagram can be derived as shown in **Figure 4.11**a. In such a band structure, electrons in $MAPbI_3$ can transfer to Nb_2CT_x, which is proved by the quenched PL intensities in **Figure 4.11**b. In return, the higher conduction band (CB: 3.9 eV) of $MAPbI_3$ (band gap: 1.5 eV) can serve as a barrier for hot carriers in Nb_2CT_x to suppress the dark current.[16] Therefore, the band-to-band absorption of $MAPbI_3$ in the visible range and plasmon-induced absorption of Nb_2CT_x in the NIR range are synergistically combined for visible-NIR photodetectors.

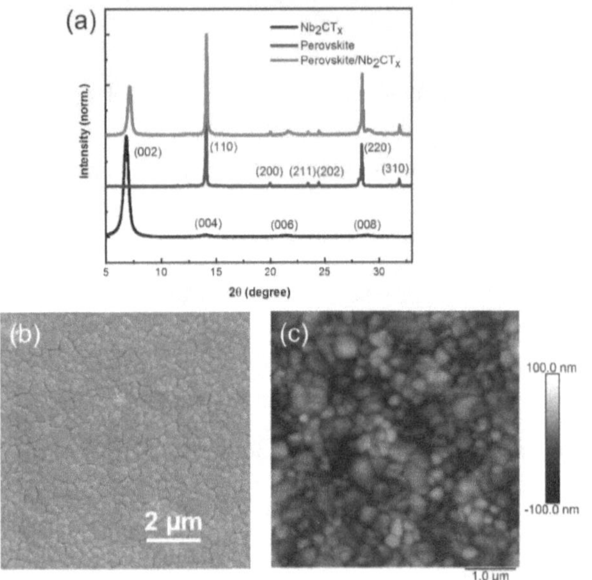

Figure 4.10 Characterizations of $MAPbI_3/Nb_2CT_x$ heterostructures. (a) XRD patterns of Nb_2CT_x, $MAPbI_3$, and $MAPbI_3/Nb_2CT_x$ heterostructures. (b) (c) SEM and AFM height images of a $MAPbI_3$ film on the Nb_2CT_x film.

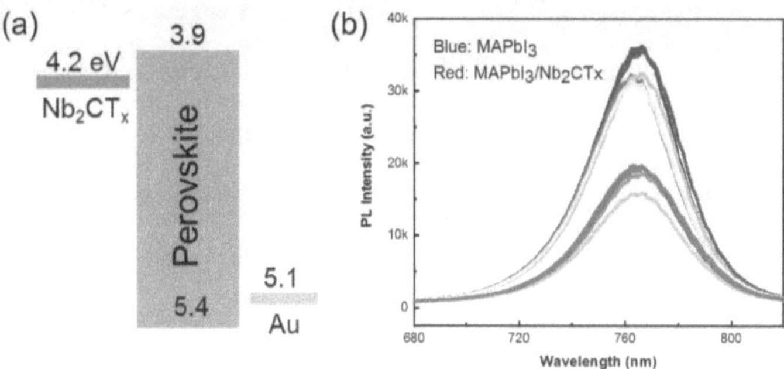

Figure 4.11 Interactions between the MAPbI$_3$/Nb$_2$CT$_x$ heterostructure. (a) Band diagram of the heterostructure for photodiodes. (b) PL spectra of perovskite films on glass substrates and Nb$_2$CT$_x$ films.

4.3.5 Photo response of Nb$_2$CT$_x$ film to NIR laser

The capability of Nb$_2$CT$_x$ to detect NIR was first examined by fabricating planar photodetectors on the Nb$_2$CT$_x$ films. The device is shown in the inset of **Figure 4.12**a with an active area of 60×1000 μm^2. The thickness of the films is around 20 nm. The performance is reflected by the current-voltage (*I-V*) curves in **Figure 4.12**a under the illumination of a 1064-nm laser. Clear increasing currents can be observed under rising laser intensities. The photocurrent (I_{ph}), which is the difference of currents under light (I_{light}) and in the dark (I_{dark}), is at the level of 10^{-5} A under a laser intensity of 34.5 mW/cm^2 and increased to 10^{-4} A level when the laser intensity reaches 62.1 mW/cm^2 as shown in **Figure 4.12**b. Figure-of-merits to evaluate the detector performance are photoresponsivity (*R*) and specific detectivity (*D**). Here we assume that the shot noise dominates the noise current. The responsivity is near 2.3 A/W. The corresponding detectivity is around 6.9×10^{9} Jones. Such high photocurrents indicate large amounts of hot carriers are induced with the plasmon relaxation in Nb_2CT_x. However, the dark current ($\sim 2.4 \times 10^{-4}$ A at 1 V) is quite high, resulting in small on/off ratios (I_{light}/I_{dark}, less than 2). Besides, the rise and delay times seen from the temporal response in **Figure 4.12**c are around 40 s and 20 s, respectively, demonstrating a slow response to the laser. The slow response may originate from the masses of defects brought by chemical etching and grain boundaries in the films. The high dark current, low on/off ratio, and slow response time significantly restrict the applications of Nb_2CT_x in practical NIR photodetection. Devices with improved performance are demanded.

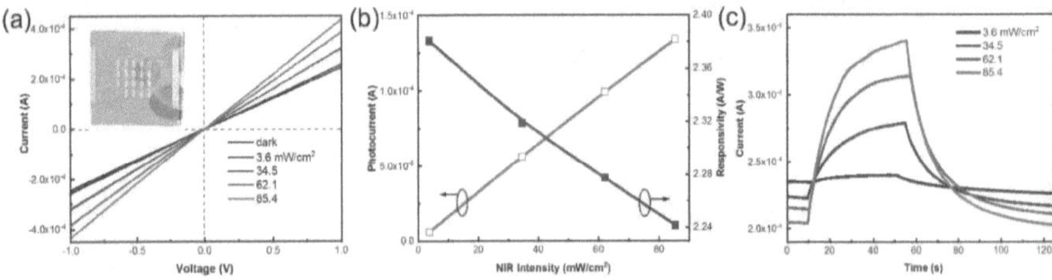

Figure 4.12 Photoresponse of Nb_2CT_x films to NIR laser. (a) Photoresponse of Nb_2CT_x under the illumination of a 1064-nm laser with different intensities. The inset in the digital photo of planar photodetectors with a 20-nm-thick Nb_2CT_x film. The length and width for each device are 70 × 1000 μm. (h) Photocurrent and responsivity of planar Nb_2CT_x photodetectors under 1064-nm laser illumination. (i) Temporal response of the device under different intensities.

4.3.6   Photoresponse of MAPbI₃/Nb₂CTₓ heterostructures

Figure 4.13a shows the schematic of the fabricated device with a MAPbI₃/Nb₂CTₓ heterostructure. In brief, the synthesized Nb_2CT_x sheet dispersion in water was first spray-coated onto glass/sapphire substrates. The MAPbI₃ layers were spin-coated onto the prepared uniform Nb_2CT_x films to form heterostructures, which is depicted in **Figure 4.13**b

for carrier transfer. The gold bottom electrode on the uncovered Nb_2CT_x and top electrodes on the perovskite surface were deposited using a shadow mask. Finally, the devices shown in the inset in **Figure 4.13**a with different areas were annealed at 100 °C for 1 h and aged for one day. It should be noticed that the annealing and aging processes are of great importance for well-functioned devices.

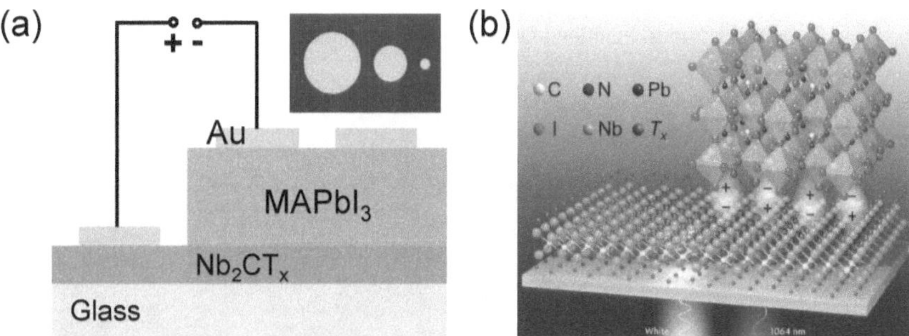

Figure 4.13 Schematic of MAPbI₃/Nb₂CTx heterostructures for photodiodes on glass substrates, Nb_2CT_x (orange), cyan (MAPbI₃). The yellow round/strip are gold electrodes on MAPbI₃/Nb₂CT_x. The bias reference is the top gold electrode. Inset is the optical microscope image of an actual device with round-top gold electrodes. The diameters are 500, 300, and 100 μm, respectively. (b) Schematic of MAPbI₃/Nb₂CT_x heterostructure for charge transfer and the plasmonic effects at the interface. H atoms and surface groups in Nb₂CT_x are not shown. The red area indicates the plasmonic effects.

Hence, the MAPbI₃ layer was brought in to establish the heterostructure with Nb₂CT_x. Vertical photodiodes were fabricated on MAPbI₃/Nb₂CT_x heterostructures to improve the detector performance and extend the detection bandwidth. The device performance in the visible range was first examined. **Figure 4.14**a shows the double swept *I-V* curves of a device in the dark and under white light with intensities from 64 nW/cm² to 2 mW/cm². The hysteresis can be observed since it exists even in single-crystal without buffer layers.[160] Clear rectification behavior can be identified in the dark current because of the built-in electric field in the $Au/MAPbI_3/Nb_2CT_x$. Therefore, open-circuit voltage (V_{oc}) is expected. For convenience, we take the values of V_{oc} from the forward scan curves here. Apparently, the photocurrent is enlarged substantially at the positive bias region with increased white light intensities. The V_{oc} exhibits the same trend as the photocurrent and the value is 0.58 V in the forward scan at a light intensity of 2 mW/cm^2. Such a V_{oc} enables the device to function in a self-powered approach without external bias.

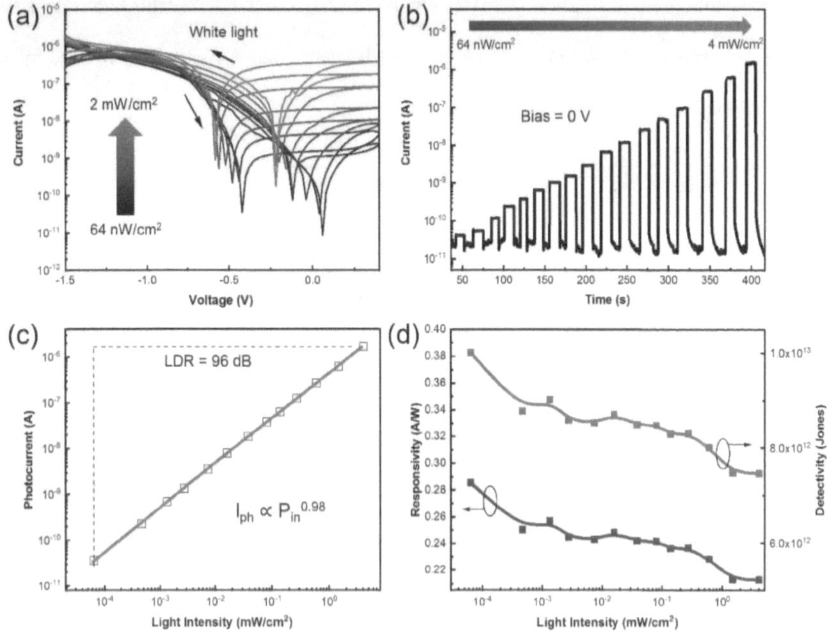

Figure 4.14 Photoresponse of MAPbI₃/Nb₂CTₓ heterostructures under white light illumination. (a) Doulbe swept *I-V* curves of the device in the dark and under white light intensities from 64 nW/cm² to 2 mW/cm². (b) Self-powered currents under white light with different intensities. (c) Plot of photocurrent versus white light intensity to show the linear response range. (d) Responsivity and detectivity of the device under different white light intensities.

Figure 4.14b shows the self-powered currents under white light with various intensities. With a zero bias, the current is around 2.5×10^{-11} A in the dark. It becomes 5.4×10^{-11} A at a light intensity of 64 nW/cm² and continuously increases to 1.6×10^{-6} A at a light intensity of 4 mW/cm², leading to an on/off ratio of around 10^5. Both the rise and decay times are less than 30 ms as shown in **Figure 4.15**, indicating a fast response of the photodetectors. The photocurrent plot versus white light intensity in **Figure 4.14**c reveals that the photocurrent increases almost linearly ($I_{ph} \propto P_{in}^{0.98}$) with light intensities in the range of 64 nW/cm^2 to 4 mW/cm^2. Such linear relationship is characterized by the linear dynamic range (LDR), which is defined as:

Equation 4.2 $$LDR = 20\log\frac{P_{max}}{P_{min}}$$

where, P_{max} and P_{min} are the maximum and minimum incident light intensities in the linear range.[161] Thus, the LDR is 96 dB, which is slightly less than reported photodetectors with buffer layers.[162, 163] In **Figure 4.14**d, it can also be observed that the responsivity (around 0.25 A/W) changes slightly in the linear response range, confirming the measured large LDR in our devices. Similarly, the detectivity is near 9×10^{12} Jones in the linear range. Therefore, the MAPbI$_3$/Nb$_2$CT$_x$ heterostructure exhibits good self-powered photodetector performance in the visible range.

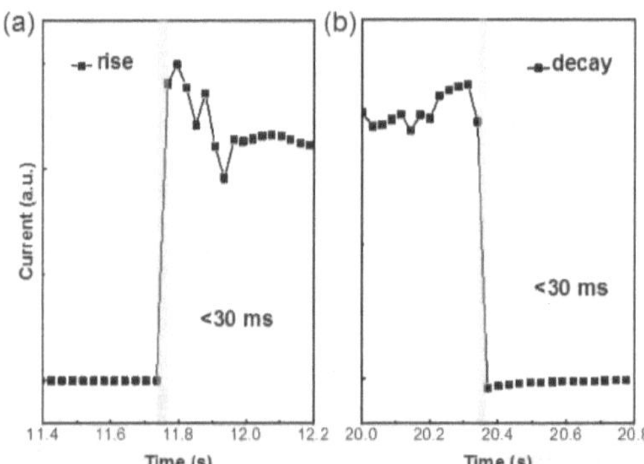

Figure 4.15 Temporal response of the device under white light illumination. (a) Rise and (b) decay times. Both rise and decay times should be less than 30 ms since 30 ms is the interval of data points and there is no more data point in the shaded (blue) area.

The detection performance in the NIR region was tested under the illumination of a 1064-nm laser. The double sweep *I-V* curve is shown in **Figure 4.16**, in which no obvious *Voc* can be observed obviously because of the small energy gap between CB of MAPbI$_3$ and Fermi level of Nb$_2$CT$_x$. The transmittace spectrum of the glass substrate is shown in **Figure 4.17**, which reveals that our glass substrates have only slight absorption (< 10%) to the 1064-nm laser. **Figure 4.18**a shows the currents of the device under the laser intensities from 1.83 mW/cm^2 to 41.75 mW/cm^2 with a zero bias, validating the self-powered detection capability in the NIR region as well. The current is around 3.6×10^{-11} A in the dark and 1.1×10^{-8} A at a laser intensity of 41.75 mW/cm^2, giving an on/off ratio of about 10^2. The device also works with a positive/negative bias ($\pm$ 0.3 V) as shown in **Figure 4.18**b. The on/off ratios are near 10^2 with higher dark and photocurrents. However, the response time increases evidently, indicating that trap/detrap processes may be involved for larger currents.[164]

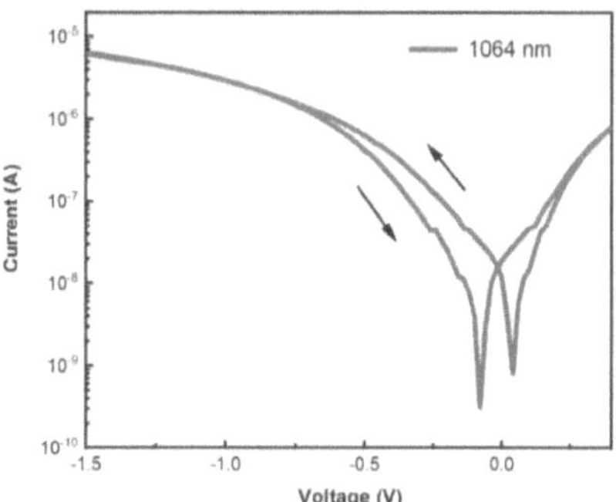

Figure 4.16 Double swept I-V curve of the device under the illumination of the 1064-nm laser at an intensity of 41.75 mW/cm^2.

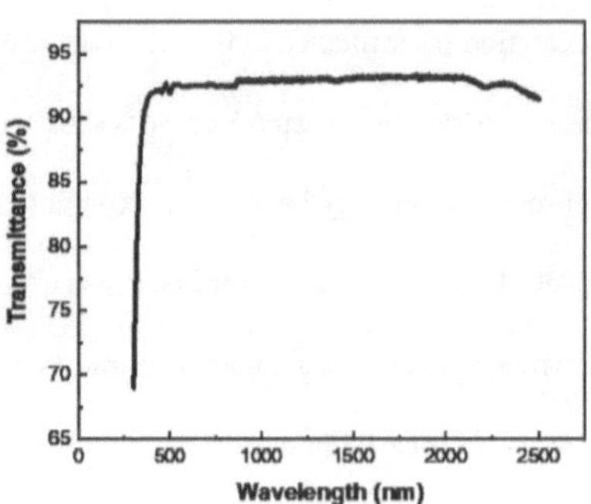

Figure 4.17 Transmittance of the glass substrate.

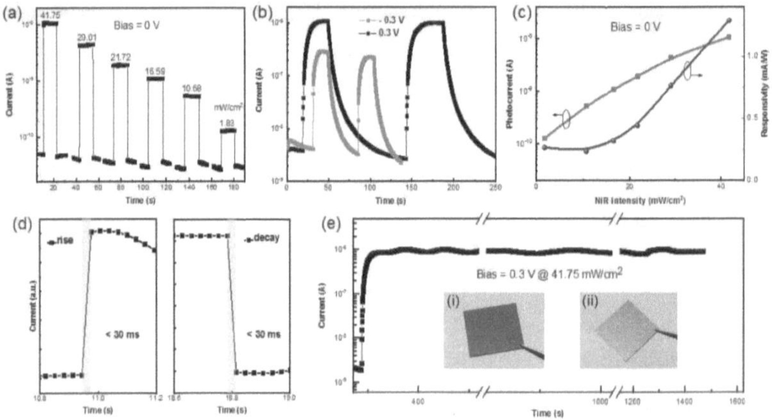

Figure 4.18 Photoresponse of MAPbI₃/Nb₂CT_x heterostructures under NIR (1064 nm) illumination. (a) Currents of the device under 1064-nm laser intensities from 1.83 to 41.75 mW/cm² with a bias of zero V. (b) Currents of the device with biases of 0.3 V (dark black) and -0.3 V (light black) under a laser intensity of 41.75 mW/cm². (c) Photocurrent and responsivity of the device with a bias of 0 V under various laser intensities. (d) Temporal dependent current to show the rise and decay time. The data point interval is 30 ms, so both the rise and decay times are less than 30 ms. (e) Photostability of the device under continuous illumination of 1064-nm laser with an intensity of 41.75 mW/cm². The bias is 0.3 V for a large photocurrent. The inset shows the perovskite film (i) before and (ii) after illuminated by the laser.

Figure 4.18c shows the photocurrents and responsivity under various laser intensities. The photocurrent and responsivity at a laser intensity of 41.75 mW/cm^2 are 1.1×10^{-8} A and 1.3 mA/W, respectively, giving the detectivity of 1.8×10^{10} Jones. The plasmon-induced charge transfer from Nb_2CT_x to $MAPbI_3$ is gauged by the quantum efficiency (QE), which is defined as:

Equation 4.3
$$QE = \frac{I_{ph}/q}{P_{in}\cdot A/E_\lambda}$$

where, E_λ is the energy of the incident photon (1064 nm).[58] The QE at the laser intensity of 41.75 mW/cm^2 is 0.015% without the bias and increases to 1.43% with a bias of 0.3 V. Such values are reasonable considering the efficiency for plasmon-induced charge transfer is usually less than 1% or even 0.01%.[154, 165-167] **Figure 4.18**d shows that both the rise and decay times of the device without bias are less than 30 ms. Therefore, although the responsivity is much smaller due to the reduced photocurrent compared to the planar Nb_2CT_x detector, the response time and the on/off ratio are promoted in a self-powered approach. Meanwhile, the device exhibits good stability in ambient conditions, even biased at 0.3 V for a large current as shown in **Figure 4.18**e. No pronounced current drop is inspected after illuminated for more than 20 min. Thus, the $MAPbI_3/Nb_2CT_x$ heterostructure presents appealing performance not only in the visible but also in the NIR region.

4.3.7 Passivated $MAPbI_3/Nb_2CT_x$ interfaces

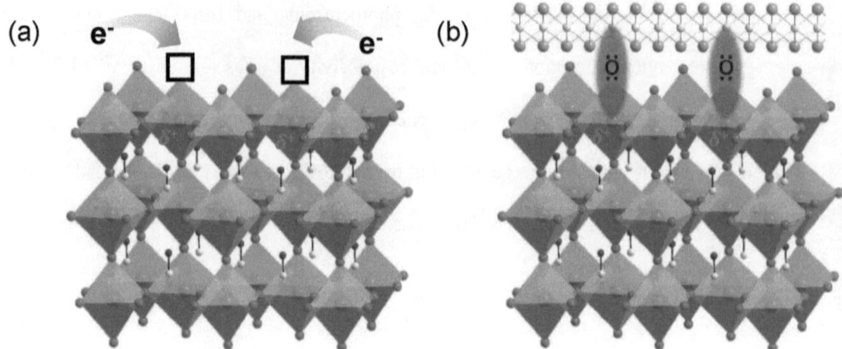

Figure 4.19 Passivation of the MAPbI₃/Nb₂CTₓ heterostructures. (a) Schematic of I vacancy sites in the MAPbI₃ surface represented by the hollow boxes. Undercoordinate Pb^{2+} ions have net positive charges and can attract electrons. (b) Pairs of electrons donated by the surface groups (T, represented by –O here) in Nb₂CTₓ sheets to neutralize positive charges in Pb^{2+} ions. Red ellipses express the coordinate bonds between Pb and T.

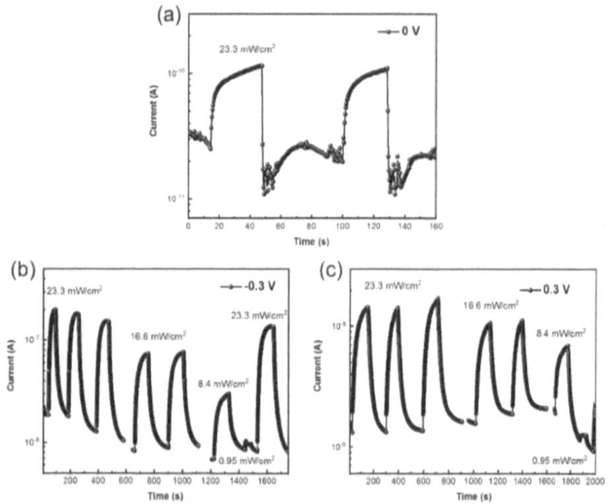

Figure 4.20 Photoresponse of as-prepared MAPbI₃/Nb₂CTₓ heterostructures without passivation.

The results above manifest that the charge, both band-to-band excited in the visible and plasmon-induced in the NIR, can transfer in the $MAPbI_3/Nb_2CT_x$ heterostructure. The successful charge transfer requires a uniform and passivated $MAPbI_3/Nb_2CT_x$ interface. A common imperfection on the surface of halide-perovskite films is the undercoordinated Pb^{2+} ions as shown in **Figure 4.19**a. Net positive charges reside on these Pb^{2+} ions and are inclined to attract electrons, enhancing the nonradiative charge recombination, diminishing carrier lifetimes, and increasing trap densities.[66, 168] A well-studied approach to passivate these imperfections is to neutralize the net positive charge and coordinate such Pb^{2+} ions by pairs of nonbonding electrons in some Lewis bases.[66, 169-171] Considering a large number of lone pair electrons in the Nb_2CT_x surface groups (T= -O, -F, -OH), it is rational to speculate that these electron pairs can coordinate the undercoordinated Pb^{2+} ions in the $MAPbI_3$ film surface to form the Pb-T bonds as shown in **Figure 4.19**b. In return, the surface groups of Nb_2CT_x are also passivated by $MAPbI_3$, reaching synergistic passivation for better charge transfer.

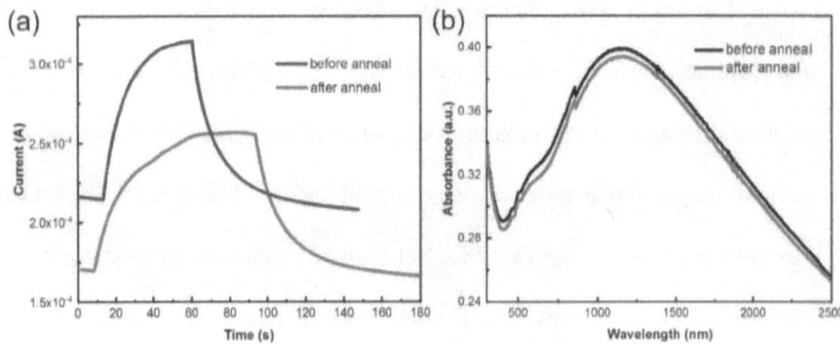

Figure 4.21 Annealing effects on the Nb$_2$CT$_x$ films. (a) Photoresponse of planar Nb$_2$CT$_x$ photodetectors under the illumination of the 1064-nm laser at an intensity of 85.4 mW/cm^2 before (blue) and after (red) annealing at 100 °C for 1 h inside the glovebox. (b) Absorbance spectra of Nb$_2$CT$_x$ films on sapphire before (black) and after (red) annealing at 100 °C for 1 h inside the glovebox. No noticeable change is observed.

Devices without annealing and aging process after the deposition of top gold electrodes were tested under the illumination of the 1064-nm laser as well to investigate the passivation of the MAPbI$_3$/Nb$_2$CT$_x$ interface. In these devices (named as-prepared samples), fewer Pb-T coordinate bonds should be formed in the MAPbI$_3$/Nb$_2$CT$_x$ interface. Indeed, the photoresponse of the as-prepared devices under different biases and laser intensities in **Figure 4.20** displays much inferior performance, including longer response times (up to 10 s at 0 V), lower on/off ratios (around 10 at 0 V), smaller photocurrents (10^{-10} A level at 0 V). It should be noted that the one-hour annealing plays no evident influence on Nb$_2$CT$_x$, which is demonstrated by the similar photoresponse of planar Nb$_2$CT$_x$-only photodetectors in **Figure 4.21**a before and after annealing of Nb$_2$CT$_x$ and the same absorption spectra in **Figure 4.21**b. Trap densities in both kinds of samples were measured using the space charge limit current (SCLC) method. The device structures for (Au/Nb$_2$CT$_x$/MAPbI$_3$/PCBM/Au) for the electron-only SCLC measurements is shown in

Figure 4.22a. For better comparison, devices with TiO_2 replacing Nb_2CT_x were also measured since the CB of TiO_2 is the same as the work function of Nb_2CT_x **Figure 4.22**b. **Figure 4.22**c shows the results of the SCLC measurements with clear Ohmic and trap-filled limit (TFL) regions. Their intersection voltage (V_{TFL}) is applied to calculate the trap density (n_t) according to the equation:

Equation 4.4 $$n_t = \frac{2\varepsilon_0\varepsilon_r V_{TFL}}{ed^2}$$

where, ε_0 is the permittivity of free space, ε_r is the relative dielectric constant of $MAPbI_3$ (32), d is the thickness of $MAPbI_3$ film (800 nm).[32] A smaller V_{TFL} for the passivated $MAPbI_3/Nb_2CTx$ heterostructure can be observed, suggesting a lower trap density (9.5×10^{14} cm^{-3}) as shown in **Figure 4.22**d. Although it is a bulk trap density, a major contribution to the reduced trap density should be from the interface, giving other same conditions.

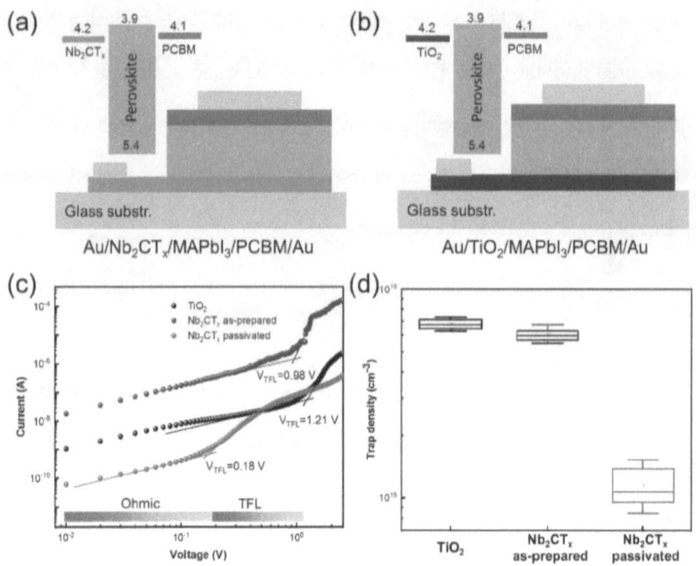

Figure 4.22 SCLC measurements. (a) (b) Device structures and band diagrams with different layers to electron-only SCLC measurements. (c) Electron-only SCLC measurements to calculated (d) Average trap densities (10 samples) in devices with TiO_2, as-prepared and passivated $MAPbI_3/Nb_2CT_x$ heterostructures.

The interfacial bonds between undercoordinated Pb^{2+} ions in $MAPbI_3$ surfaces and surface groups of Nb_2CT_x can be reflected from capacitance directly. **Figure 4.23**a exhibits the measured frequency (20- 5×10^4 Hz) dependent capacitances for the as-prepared and passivated $MAPbI_3/Nb_2CTx$ heterostructures. At low-frequencies ($< 10^3$ Hz), the capacitance in the as-prepared sample is much larger than that of the passivated one (16 times higher at 20 Hz). In intermediate frequencies (10^3-10^4 Hz), their capacitance is almost the same. At higher frequencies, both capacitances decrease with different slopes related to serious resistances.[172] The higher capacitance at low frequencies is attributed to the excess capacitance from the $MAPbI_3/Nb_2CT_x$ interface states since the interface state can only follow the alternative signal at low frequencies ($< 10^3$ Hz).[173, 174] In addition, the frequency-dependent conductance in **Figure 4.24** manifests a larger conductance for the as-prepared one, resulting from flips of interface states. **Figure 4.23**b presents the calculated relative dielectric constants using a parallel-plate capacitor model. A value of 31 is obtained for the passivated sample at 20 Hz, which is close to the reported value (32).[32] Such results firmly support that the interface state density in the passivated $MAPbI_3/Nb_2CT_x$ drops enormously due to the Pb-T bonds.

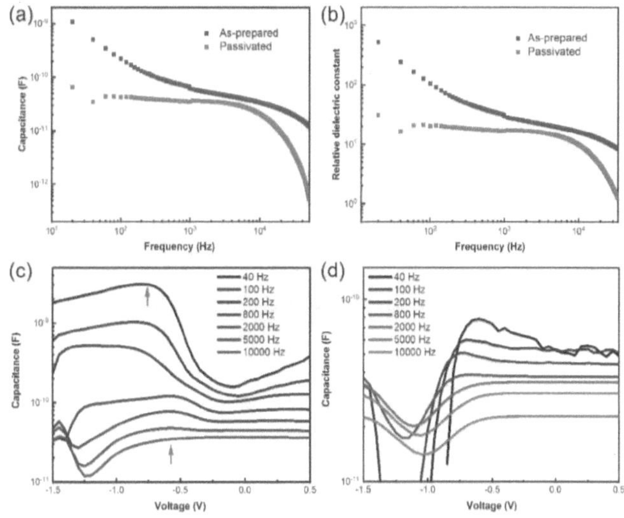

Figure 4.23 Capacitance of the $MAPbI_3/Nb_2CT_x$ heterostructures in the dark. The measured devices have the same structures as their corresponding photodetectors. (a) (b) Plots of capacitance and relative dielectric constant versus frequency for as-prepared and passivated $MAPbI_3/Nb_2CT_x$ heterostructures, respectively. (c) (d) Plots of capacitance versus voltage for as-prepared and passivated $MAPbI_3/Nb_2CT_x$ heterostructures, respectively. Red arrows in (c) indicate the peak positions at low and high frequencies.

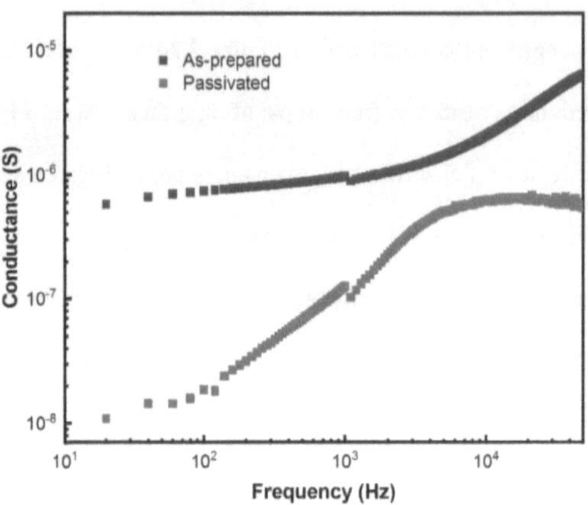

Figure 4.24 Frequency-dependent conductance of the as-prepared and passivated MAPbI₃/Nb₂CTₓ heterostructures.

The capacitance-voltage curves with various frequencies were also measured for the as-prepared and passivated samples as shown in **Figure 4.23**c-d. At low frequencies, the capacitance for the as-prepared sample is much higher, which matches the results in Figure 6a. A distinct peak can be discovered in both samples, which is typical in Schottky diodes and whose voltage and height are related to the built-in electric field, electrode Fermi level, temperature, and interface state density.[172, 175] In the as-prepared sample, the peak voltage at 40 Hz is around -0.75 V. It becomes -0.58 V when the frequency is larger than 2×10^3 Hz, which is almost the same as that in the passivated one regardless of frequencies. The peak value of -0.58 V is also in accordance with the V_{oc} in **Figure 4.14**a. The reduced peak voltages and decreased peak height in the passivated sample at low frequencies demonstrate a lower interface state density in the passivated sample.

Furthermore, the capacitance is increasing with voltage (0- 0.5 V) at low frequencies in the as-prepared sample, which is not seen in the passivated sample and can be ascribed to increasing alignments of surface groups in Nb_2CT_x. Hence, it can be deduced that the surface groups in Nb_2CT_x are bonded with undercoordinated Pb^{2+} in $MAPbI_3$.

In order to detect the Pb-O bond at the interface, we perfromed Raman analsysi. The results of the as-prepared (pristine) $MAPbI_3$and passivated heterostructure film are shown in **Figure 4.25**. The Raman spectrum of pristine $MAPbI_3$ is similar to previous reports, in which the 83 cm^{-1}, 107cm^{-1}, and 122 cm^{-1} peaks are related to bonds involving Pb atoms.[176-178] We can also see two other peaks at 152 cm^{-1} and 176 cm^{-1}. The peak corresponding to Pb-O bonds in PbO is located at around 142 cm^{-1}.[179] Therefore, we believe that the new peaks are attributed to Pb-O bonds, but likely with weaker bonding at the interface compared with bulk PbO.

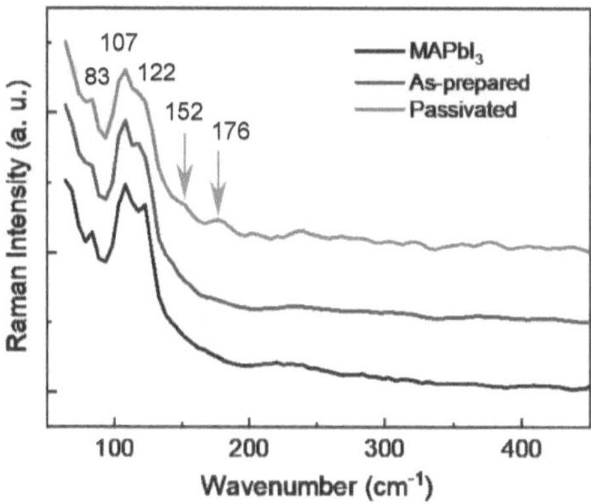

Figure 4.25 Raman spectra of as-prepared (pristine) $MAPbI_3$and passivated heterostructure films.

4.4 Conclusion

In conclusion, self-powered photodiodes functioned in the visible-NIR region are prepared on $MAPbI_3/Nb_2CT_x$ heterostructures. The combination of band-to-band absorption of $MAPbI_3$ in the visible and plasmonic absorption of Nb_2CT_x in the NIR render the detector ability to detect both visible and NIR light. The matched band diagram can also suppress the dark current of Nb_2CT_x in the NIR region. As a consequence, the detector shows a responsivity of 0.25 A/W and an LDR of 96 dB under white light intensities from $64\ nW/cm^2$ to $4\ mW/cm^2$. In the NIR region (1064 nm), the device exhibits an on/off ratio of around 10^2 and response times of less than 30 ms. Both enhance hugely compared to the planar Nb_2CT_x-only detector. The improved performance is attributed to the efficient and rapid charge transfer in the passivated $MAPbI_3/Nb_2CT_x$ interface. The passivation results from the coordinate bonding between undercoordinate Pb^{2+} ions in $MAPbI_3$ surfaces and surface groups in Nb_2CT_x, demonstrated by the SCLC and capacitance measurements. This work provides a new platform for investigating perovskite/MXene interfaces and stimulates further research on optoelectronic applications of MXenes.

Chapter 5 Summary and outlook

This chapter gives a summary of the mixed-dimensional heterostructure systems we investigated with metal-halide perovskites and different types of 2D materials, including semimetallic monolayer graphene, semiconducting phosphorus-doped graphitic carbon nitride, and plasmonic Nb_2CT_x MXene. The interface interactions between metal-halide perovskites and these 2D materials and optoelectronic devices based on these heterostructures will be concluded. These perovskite-based heterostructures offer unique optical and optoelectronic properties, imbuing them with new functionality and better performance of optoelectronic devices. Besides, we will discuss the potential research directions of mixed-dimensional heterostructures based on metal-halide perovskites and 2D materials. We believe the growth of large-scale, high-quality, thin single-crystalline perovskite films on 2D materials is the key to practical applications. Moreover, the patterning of the heterostructures is of great importance for the miniaturization and integration of devices. At last, characteristic interactions between different perovskites and 2D materials should be surveyed and exploited for desired functional electronic and optoelectronic devices. We think that the investigation of heterostructures based on perovskites and newly developed 2D materials like MXenes should be a good direction.

5.1 Summary

In order to improve the performance and explore new functionality of halide perovskite devices, we prepared heterostructures with different 2D materials including semi-metallic graphene, semiconducting phosphorus-doped graphitic carbon nitride, and plasmonic niobium carbide MXene. Then optoelectronic properties were investigated and photodetector devices were fabricated using these heterostructures. We first prepared the vdW mixed-dimensional single-crystalline $MAPbBr_3$/graphene heterostructure by selective growth of 3D $MAPbBr_3$ platelets on patterned 2D single-layer graphene using a one-step CVD method. Compared to the previous two-step conversion synthesis of perovskites, the one-step growth can make the vdW heterostructure less defects for intrinsic investigations of properties like charge transfer on the interface. Interestingly, perovskite platelets show preferred growth on graphene surfaces. The charge transfer is observed from the significant photoluminescence quenching. Besides, p-type doping in the graphene layer is revealed by Raman spectra with a Fermi level decrease of 272 meV in graphene, corresponding to an amount of 7.5×10^{12} cm^{-2}. And negatively charged surfaces caused by electrons remaining in the thin perovskite platelets are evidenced by the reduced surface potentials measured by the KPFM. The upward band bending is reflected by the current-voltage curves recorded by CAFM. Moreover, field-effect phototransistors fabricated using the perovskite/graphene heterostructure channels show increased Dirac voltages under illumination, suggesting an enhanced p-type character in graphene. These results strongly prove that charge transfer occurs between $MAPbBr_3$ and graphene.

The charge transfer between halide perovskite and 2D materials can be used to improved the performance like photodetectorsThen we fabricated high-performance and air-stable broadband photodetectors made of MLHP and 2D PCN-S. Graphitic carbon nitride was chosen because of its easy synthesis, metal-free composition, and low cost. Furthermore, the band structure of PCN-S is tuned by the doping of phosphorus and demonstrated to match with MLHP to assist the charge transfer evidenced by the quenched photoluminescence intensity and reduced lifetime. The as-fabricated PCN-S/MLHP hybrid photodetectors from solution show increased photocurrent and reduced dark current, leading to two orders of magnitude higher on/off ratio of 10^5, one order of magnitude higher photodetectivity of 10^{13} *Jones* and responsivity of 14 A W^{-1} with reliable and fast photoswitching characteristics. In addition, the hybrid photodetectors exhibit increased moisture-resistance thanks to the improved surface hydrophobicity seen from the larger contact angle with water. Our results show a new approach to fabricate perovskite photodetectors with improved and more stable performance by creating bulk heterojunctions with 2D materials.

We further try to expand the detection band of the perovskite photodetectors since most of them work in UV-visible ranges. Therefore, we fabricated vertical photodiodes using $MAPbI_3$ and Nb_2CT_x MXene heterostructures. We can use the band to band absorption of halide perovskite to detect white light and the surface plasmonic absorption of Nb_2CT_x to detect NIR lasers. $MAPbI_3$ and Nb_2CT_x have a matching band structure and the heterostructure is exploited for self-powered visible-NIR photodiodes. The use of the $MAPbI_3$ layer expands the operation of the diode to the visible range while suppressing the dark current of the NIR-absorbing Nb_2CT_x layer. As a result, the photodiode responds linearly under white light illumination with a responsivity of 0.25 A/W. Furthermore, when illuminated with a 1064-nm laser, the photodiode demonstrates a higher on/off ratio ($\sim 10^2$) and faster response times (< 30 ms) compared to that of planar Nb_2CT_x-only detectors ($<$ 2 and 20 s, respectively). Experiments from space-charge limited current and capacitance measurements show that the coordinate bonding between the surface groups of the MXene and the undercoordinated Pb^{2+} ions in the $MAPbI_3$ surface has led to a passivated $MAPbI_3/Nb_2CT_x$ interface resulting in an efficient and speeded charge transfer.

Overall, our results show that heterostructures offer a platform to design new devices with improved performance.

5.2 Growth of high-quality perovskite film

The large-scale preparation of high-quality perovskite films is regarded as a crucial step to fabricate the mixed-dimensional heterostructures. A high-quality perovskite film is the guarantee of the successful preparation of heterostructures and the accurate investigations of properties and devices. Therefore, a fundamental direction is to grow large-scale, thin, and single-crystalline perovskite to replacing films with one or several imperfections like small-size, thick, and poly-crystalline as shown in **Figure 5.1**a. So far, intense efforts have been engaged in synthesizing large-scale single-crystal thin films. For example, Wei et al. prepared large-scale, single-crystalline $MAPbBr_3$ films (1 μm) from a solution method as shown in **Figure 5.1**b. Also, Jie et al. synthesized large-scale, single-crystalline $CsPbBr_3$ films (1 μm) from the CVD method **Figure 5.1**c. These films show better optoelectronic and electric properties compared to their counterparts, which are polycrystal films here. However, the thickness is large, reaching micrometer-level thicknesses. Such large thicknesses can exceed the carrier diffusion lengths, outpace the junction width, and diminish the field effect, which leads to the poor regulation of properties and low performance of devices. Therefore, besides large-scale, single-crystalline, being thin should be another concern to be high-quality films for heterostructures. To achieve high-quality films on demanded substrates, growth conditions and surface properties like hydrophilicity and hydrophobicity can be tuned. Our experiments have shown that it is possible to grow large-scale and thin CsPbBr$_3$ films (less than 100 nm) on mica substrates by adjusting the growth temperature and pressure in the CVD method as shown in **Figure 5.2**a. However, aligned grain boundaries still exist in the films with the perovskite grains in sizes of a few micrometers as shown in **Figure 5.2**b. Characterizations like XRD patterns and PL spectrum in **Figure 5.2**c-d show the successful growth and well-aligned growth direction and textures. The fabricated photodetectors reveal good photoresponses as shown in **Figure 5.2**e-f. Therefore, it is possible to grow large-scale, thin, single-crystalline perovskite films if all parameters are further optimized.

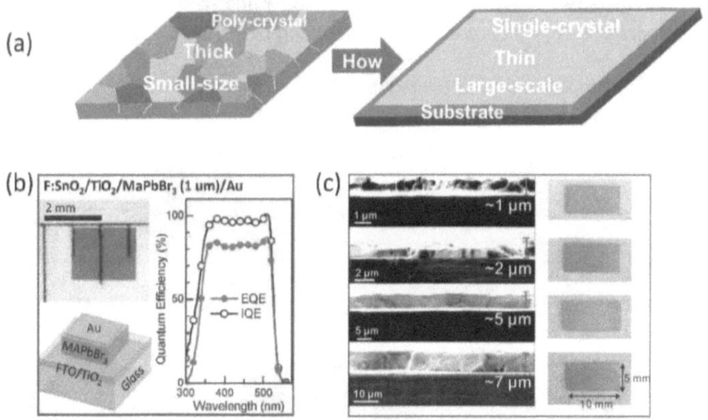

Figure 5.1 Preaperation of high-quality metal-halide perovksite films. (a) Requirements of high-quality metal-halide perovskite films. (b) Large-scale thick MAPbBr$_3$ single-crystalline films prepared from a solution method.[180] (c) Large-scale thick CsPbBr$_3$ single-crystalline films prepared from a vapor method.[181]

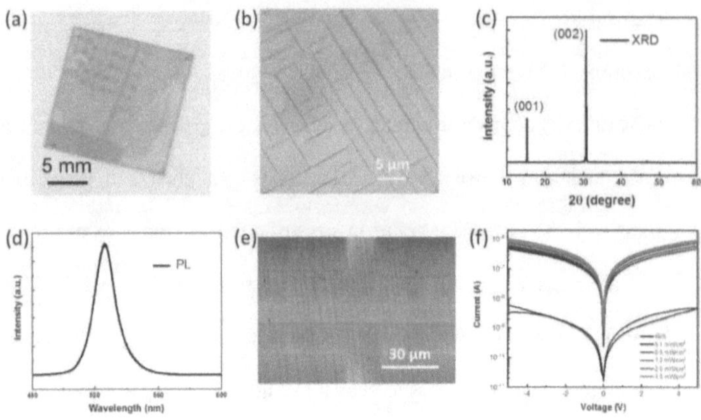

Figure 5.2 CsPbBr$_3$ film on mica and other substrates. (a) (b) Digital and optical microscopy images of CsPbBr$_3$ film on mica, respectively. (c) XRD patterns. (d) PL spectrum. (e) (f) Photodetector device and performance, respectively.

5.3 Patterning of heterostructures

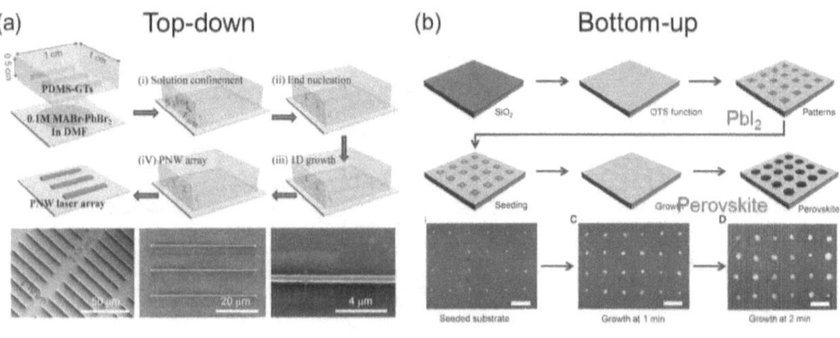

Figure 5.3 Patterning of metal-halide perovskites. (a) Top-down micro-pattern method using PDMS.[182] (b) Bottom-up selective growth method by modifying the surface wettability.[183]

It is very impactful if we can achieve patterning of these heterostructures on selected substrates. The patterning can be accomplished either by top-down photolithography or bottom-up selective growths. For metal-halide perovskites, their instability to polar solvents like acetone and water makes it hard to use normal photolithography processes and solvents. Thus, special techniques are usually demanded for the photolithograph of metal-halide perovskites.[184] The so-called micro-pattern method is another option, in which patterns can be fabricated on polydimethylsiloxane (PDMS) first and then transferred to the perovskite as shown in **Figure 5.3a**.[182] Therefore, the top-down method needs particular conditions and extra efforts for patterning, although it could be a general method. Besides, bottom-up methods like selective growth on SiO_2 substrates by modifying the surface wettability can also be achieved as shown in **Figure 5.3b**.[183] In this method, the substrate are patterned to hydrophilic and hydrophobic regions and the perovskites tend to grow on the former one. In our work, we also achieved selective growth of $MAPbBr_3$ platelets on patterned graphene. However, the bottom-up growth is not a general method and each case needs its own techniques. For example, it is different and hard to make MoS_2 surface hydrophilic without introducing damage compared to SiO_2. We think that mature techniques should be explored in different materials, which needs tremendous efforts.

5.4 More types of 2D materials

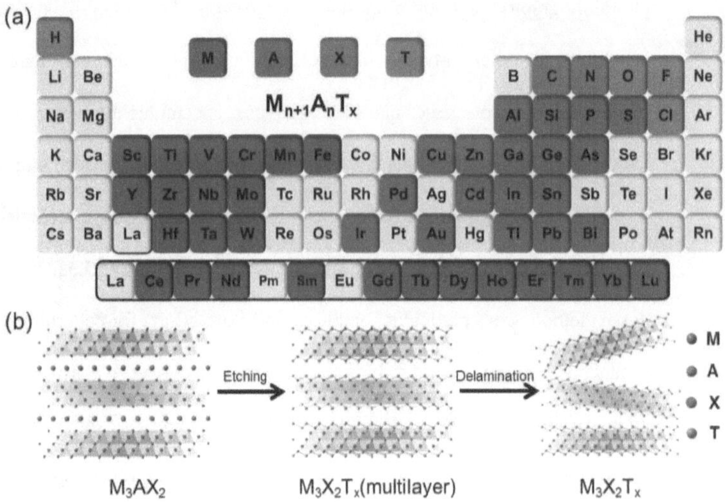

Figure 5.4 MXene spices and synthesis processes. (a) MAX and MXene spice by changing different elements. (b) Synthesis of MXenes from MAX phases using $M_3X_2T_x$ as an example.

At last, we believe it is significant to prepare metal-halide perovskites on more types of 2D materials. There are enormous 2D materials reported. Some of them have special properties and can be formed potential heterostructures with perovskites with distinct functionalization. For example, In_2Se_3 is ferroelectric, and it may have unique carrier tunability. In recent years, a new class of 2D materials named MXenes has attracted increasing attention. MXenes have the general chemical formula $M_{n+1}X_nT_x$, in which M stands for early transition metals, X can be carbon or nitrogen, T represents surface terminated groups like -O, -OH, -F and -Cl, and n = 1 − 4.[144, 185] More than 20 different MXenes have been experimentally synthesized until now and even more are predicted as shown in **Figure 5.4**a. They are thin 2D sheets with hydrophilic surface groups due to their synthesis process shown in **Figure 5.4**b. The large amount of MXenes renders wide ranges of properties like surface tunability, work function, surface plasmon, capacitance, and conductivity. Consequently, it gives many combinations for heterostructures with perovskites. In our work, we proved that the plasmonic Nb_2CT_x and $MAPbI_3$ heterostructures for visible-NIR photodiodes. We believe more MXenes should be tried, and it is promising to find suitable combinations with desired properties for high-performance devices.

5.5   New designs of devices

As mentioned in Chapter 1, the developing history of metal-halide perovskite-based electronic and optoelectronics demonstrates that they can be applied in many types of devices. Although we have made some progress in improving the performance and broadening the operation band of photodetectors, more efforts and researches are desired to make new designs of devices with specific functionalities. For example, artificial synapses are gaining many research efforts as they can mimic human neuron.[186, 187] The neuromorphic networks built with these elementary blocks possess abilities of perception, memory, and learning. Among them, optoelectronic synapses own the ability of visual perception since their photoresponse capability, which means the light stimuli can be converted to the electrical signal.[186] As a kind of extremely photo-sensitive material, metal-halide perovskites show great potential in optoelectronic artificial synapses. From the angle of our view, a possible way to apply perovskites is to form heterostructures with 2D materials as shown in **Figure 5.5**. Since we can choose either metallic or semiconducting 2D materials, Schottky or P-N junctions can be formed between perovskites and 2D materials. Charges can be trapped due to bent bands and defects. Since the band bending and charge density can be tuned by external biases and light intensities, respectively, optoelectronic artificial synapses can be achieved.

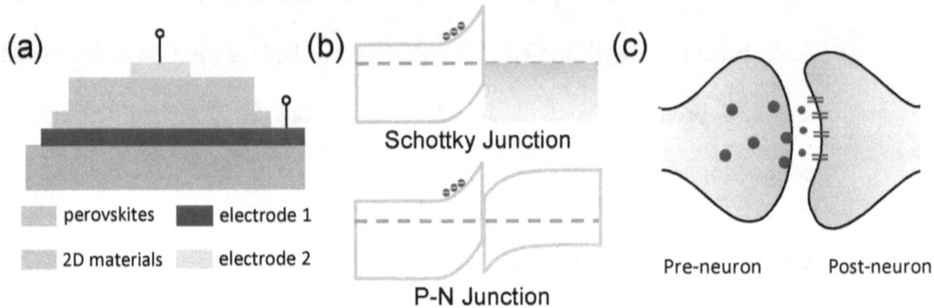

Figure 5.5 Peerovskite/2D material heterostructures applied in devices like optoelectronic synapses. (a) Possible device structure for the heterostructure. (b) Schottky and P-N junctions can be formed between perovskites and 2D materials. The purple spheres are trapped electrons due to the band bending and defects. (c) Schematic of a synapse including pre-neuron and post-neuron mimicked by the heterostructure.

5.6   Integration of devices

In real-life applications, each practical electronic device is made of many elementary ones, and some can even reach billions. In this way, those devices can possess abilities including form data capture, data storage, data analysis, and data display as shown in **Figure 5.6**. Therefore, the integration of elementary devices is essential to fabricate such complicate devices used in real life. Regarding metal-halide perovskites, they can be used in many different devices. However, these devices with a specific functionality are not integrated, which is not in favor of practically used devices. A likely reason is the patterning limitation as discussed above. Nevertheless, it is still important to integrate these devices with a low elementary device density. Previous works have demonstrated the integration of transistor-LED and photodetector-memory using organic materials and oxides,[188, 189] respectively. Therefore, it is necessary to try to integrate all metal-halide perovskite electronic and optoelectronic devices.

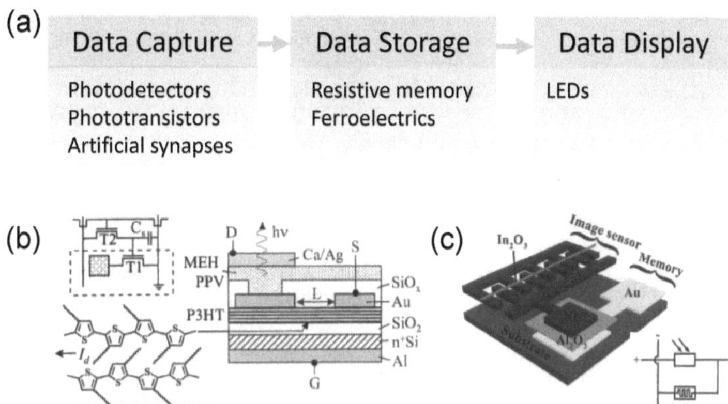

Figure 5.6 Integration of elementary devices. (a) Capabilities of practically used devices. (b) (c) Examples of transistor-LED and photodetector-memory integrations, respectively.[189]

References

(1)   Novoselov K. S.; Geim A. K.; Morozov S. V.; Jiang D.; Zhang Y.; Dubonos S. V.; Grigorieva I. V.; Firsov A. A., Electric Field Effect in Atomically Thin Carbon Films. *Science* **2004**, *306*, 666.

(2)   Carvalho A.; Wang M.; Zhu X.; Rodin A. S.; Su H.; Castro Neto A. H., Phosphorene: From Theory to Applications. *Nat. Rev. Mater.* **2016**, *1*, 16061.

(3)   Wang Y.; Qiu G.; Wang R.; Huang S.; Wang Q.; Liu Y.; Du Y.; Goddard W. A.; Kim M. J.; Xu X.; Ye P. D.; Wu W., Field-Effect Transistors Made from Solution-Grown Two-Dimensional Tellurene. *Nat. Electron.* **2018**, *1*, 228-236.

(4)   Del Alamo J. A., Nanometre-Scale Electronics with Iii–V Compound Semiconductors. *Nature* **2011**, *479*, 317-323.

(5)   Özgür Ü.; Alivov Y. I.; Liu C.; Teke A.; Reshchikov M. A.; Doğan S.; Avrutin V.; Cho S. J.; Morkoç H., A Comprehensive Review of Zno Materials and Devices. *J. Appl. Phys.* **2005**, *98*, 041301.

(6)   She X.; Huang A. Q.; Ó L.; Ozpineci B., Review of Silicon Carbide Power Devices and Their Applications. *IEEE Transactions on Industrial Electronics* **2017**, *64*, 8193-8205.

(7)   Regmi G.; Ashok A.; Chawla P.; Semalti P.; Velumani S.; Sharma S. N.; Castaneda H., Perspectives of Chalcopyrite-Based Cigse Thin-Film Solar Cell: A Review. *J. Mater. Sci.: Mater. Electron.* **2020**, *31*, 7286-7314.

(8)   Pamplin B. R.; Kiyosawa T.; Masumoto K., Ternary Chalcopyrite Compounds. *Progress in Crystal Growth and Characterization* **1979**, *1*, 331-387.

(9)   Manzeli S.; Ovchinnikov D.; Pasquier D.; Yazyev O. V.; Kis A., 2d Transition Metal Dichalcogenides. *Nat. Rev. Mater.* **2017**, *2*, 17033.

(10)   Xu X.; Das G.; He X.; Hedhili M. N.; Fabrizio E. D.; Zhang X.; Alshareef H. N., High-Performance Monolayer MoS_2 Films at the Wafer Scale by Two-Step Growth. *Adv. Funct. Mater.* **2019**, *29*, 1901070.

(11)   Forrest S. R.; Thompson M. E., Introduction: Organic Electronics and Optoelectronics. *Chem. Rev.* **2007**, *107*, 923-925.

(12)   Sekitani T.; Someya T., Stretchable, Large-Area Organic Electronics. *Adv. Mater.* **2010**, *22*, 2228-2246.

(13)   Kojima A.; Teshima K.; Shirai Y.; Miyasaka T., Organometal Halide Perovskites as Visible-Light Sensitizers for Photovoltaic Cells. *J. Am. Chem. Soc.* **2009**, *131*, 6050-6051.

(14)   Best Research-Cell Efficiency Chart. https://www.nrel.gov/pv/cell-efficiency.html (accessed April 04, 2021).

(15)   Wang K.; Yang D.; Wu C.; Sanghadasa M.; Priya S., Recent Progress in Fundamental Understanding of Halide Perovskite Semiconductors. *Progress in Materials Science* **2019,** *106*, 100580.

(16)   Haque M. A.; Sheikh A. D.; Guan X.; Wu T., Metal Oxides as Efficient Charge Transporters in Perovskite Solar Cells. *Adv. Energy Mater.* **2017,** *7*, 1602803.

(17)   Brivio F.; Frost J. M.; Skelton J. M.; Jackson A. J.; Weber O. J.; Weller M. T.; Goñi A. R.; Leguy A. M. A.; Barnes P. R. F.; Walsh A., Lattice Dynamics and Vibrational Spectra of the Orthorhombic, Tetragonal, and Cubic Phases of Methylammonium Lead Iodide. *Phys. Rev. B* **2015,** *92*, 144308.

(18)   Lin Y. H.; Pattanasattayavong P.; Anthopoulos T. D., Metal-Halide Perovskite Transistors for Printed Electronics: Challenges and Opportunities. *Adv. Mater.* **2017,** *29*.

(19)   Saparov B.; Mitzi D. B., Organic–Inorganic Perovskites: Structural Versatility for Functional Materials Design. *Chem. Rev.* **2016,** *116*, 4558-4596.

(20)   Saparov B.; Mitzi D. B., Organic-Inorganic Perovskites: Structural Versatility for Functional Materials Design. *Chem. Rev.* **2016,** *116*, 4558-96.

(21)   Karlin K. D.; Mitzi D. B., *Synthesis, Structure, and Properties of Organic-Inorganic Perovskites and Related Materials*. John Wiley & Sons, Inc.: 2007; Vol. 48.

(22)   Ahmad S.; Guo X., Rapid Development in Two-Dimensional Layered Perovskite Materials and Their Application in Solar Cells. *Chin. Chem. Lett.* **2017.**

(23)   Im J.-H.; Kim H.-S.; Park N.-G., Morphology-Photovoltaic Property Correlation in Perovskite Solar Cells: One-Step Versus Two-Step Deposition of $Ch_3nh_3pbi_3$. *APL Materials* **2014,** *2*, 081510.

(24)   Chen Q.; Zhou H.; Hong Z.; Luo S.; Duan H.-S.; Wang H.-H.; Liu Y.; Li G.; Yang Y., Planar Heterojunction Perovskite Solar Cells Via Vapor-Assisted Solution Process. *J. Am. Chem. Soc.* **2014,** *136*, 622-625.

(25)   Liu M.; Johnston M. B.; Snaith H. J., Efficient Planar Heterojunction Perovskite Solar Cells by Vapour Deposition. *Nature* **2013,** *501*, 395-398.

(26)   Ma C.; Shi Y.; Hu W.; Chiu M.-H.; Liu Z.; Bera A.; Li F.; Wang H.; Li L.-J.; Wu T., Heterostructured $Ws_2/Ch_3nh_3pbi_3$ Photoconductors with Suppressed Dark Current and Enhanced Photodetectivity. *Adv. Mater.* **2016,** *28*, 3683-3689.

(27)   Shi D.; Adinolfi V.; Comin R.; Yuan M.; Alarousu E.; Buin A.; Chen Y.; Hoogland S.; Rothenberger A.; Katsiev K.; Losovyj Y.; Zhang X.; Dowben P. A.;

Mohammed O. F.; Sargent E. H.; Bakr O. M., Low Trap-State Density and Long Carrier Diffusion in Organolead Trihalide Perovskite Single Crystals. *Science* **2015**, *347*, 519.

(28)   Kovalenko M. V.; Protesescu L.; Bodnarchuk M. I., Properties and Potential Optoelectronic Applications of Lead Halide Perovskite Nanocrystals. *Science* **2017**, *358*, 745.

(29)   Mi Y.; Liu Z.; Shang Q.; Niu X.; Shi J.; Zhang S.; Chen J.; Du W.; Wu Z.; Wang R.; Qiu X.; Hu X.; Zhang Q.; Wu T.; Liu X., Fabry–Pérot Oscillation and Room Temperature Lasing in Perovskite Cube-Corner Pyramid Cavities. *Small* **2018**, *14*, 1703136.

(30)   Wang Y.; Shi Y.; Xin G.; Lian J.; Shi J., Two-Dimensional Van Der Waals Epitaxy Kinetics in a Three-Dimensional Perovskite Halide. *Cryst. Growth Des.* **2015**, *15*, 4741-4749.

(31)   Zhu H.; Fu Y.; Meng F.; Wu X.; Gong Z.; Ding Q.; Gustafsson M. V.; Trinh M. T.; Jin S.; Zhu X. Y., Lead Halide Perovskite Nanowire Lasers with Low Lasing Thresholds and High Quality Factors. *Nat. Mater.* **2015**, *14*, 636-642.

(32)   Dong Q.; Fang Y.; Shao Y.; Mulligan P.; Qiu J.; Cao L.; Huang J., Electron-Hole Diffusion Lengths > 175 Mm in Solution-Grown $Ch_3nh_3pbi_3$ Single Crystals. *Science* **2015**, *347*, 967.

(33)   Liu Y.; Yang Z.; Cui D.; Ren X.; Sun J.; Liu X.; Zhang J.; Wei Q.; Fan H.; Yu F.; Zhang X.; Zhao C.; Liu S., Two-Inch-Sized Perovskite $Ch_3nh_3pbx_3$ (X = Cl, Br, I) Crystals: Growth and Characterization. *Adv. Mater.* **2015**, *27*, 5176-5183.

(34)   Li M.; Li B.; Cao G.; Tian J., Monolithic Mapbi3 Films for High-Efficiency Solar Cells Via Coordination and a Heat Assisted Process. *J. Mater. Chem. A* **2017**, *5*, 21313-21319.

(35)   Liu Z.; Qiu L.; Juarez-Perez E. J.; Hawash Z.; Kim T.; Jiang Y.; Wu Z.; Raga S. R.; Ono L. K.; Liu S.; Qi Y., Gas-Solid Reaction Based over One-Micrometer Thick Stable Perovskite Films for Efficient Solar Cells and Modules. *Nat. Commun.* **2018**, *9*, 3880.

(36)   Zhang Y.; Liu J.; Wang Z.; Xue Y.; Ou Q.; Polavarapu L.; Zheng J.; Qi X.; Bao Q., Synthesis, Properties, and Optical Applications of Low-Dimensional Perovskites. *Chem. Commun. (Cambridge, U. K.)* **2016**, *52*, 13637-13655.

(37)   Liu Z.; Mi Y.; Guan X.; Su Z.; Liu X.; Wu T., Morphology-Tailored Halide Perovskite Platelets and Wires: From Synthesis, Properties to Optoelectronic Devices. *Advanced Optical Materials* **2018**, *6*.

(38)   Liu Z.; Li Y.; Guan X.; Mi Y.; Al-Hussain A.; Ha S. T.; Chiu M. H.; Ma C.; Amer M. R.; Li L. J.; Liu J.; Xiong Q.; Wang J.; Liu X.; Wu T., One-Step Vapor-Phase

Synthesis and Quantum-Confined Exciton in Single-Crystal Platelets of Hybrid Halide Perovskites. *J Phys Chem Lett* **2019**, *10*, 2363-2371.

(39)   Protesescu L.; Yakunin S.; Bodnarchuk M. I.; Krieg F.; Caputo R.; Hendon C. H.; Yang R. X.; Walsh A.; Kovalenko M. V., Nanocrystals of Cesium Lead Halide Perovskites (Cspbx$_3$, X = Cl, Br, and I): Novel Optoelectronic Materials Showing Bright Emission with Wide Color Gamut. *Nano Lett.* **2015**, *15*, 3692-3696.

(40)   Green M. A.; Ho-Baillie A.; Snaith H. J., The Emergence of Perovskite Solar Cells. *Nat. Photonics* **2014**, *8*, 506-514.

(41)   Maculan G.; Sheikh A. D.; Abdelhady A. L.; Saidaminov M. I.; Haque M. A.; Murali B.; Alarousu E.; Mohammed O. F.; Wu T.; Bakr O. M., Ch3nh3pbcl3 Single Crystals: Inverse Temperature Crystallization and Visible-Blind Uv-Photodetector. *The Journal of Physical Chemistry Letters* **2015**, *6*, 3781-3786.

(42)   Yang D.; Xie C.; Xu X.; You P.; Yan F.; Yu S. F., Lasing Characteristics of Ch3nh3pbcl3 Single-Crystal Microcavities under Multiphoton Excitation. *Advanced Optical Materials* **2018**, *6*, 1700992-n/a.

(43)   Yamada T.; Aharen T.; Kanemitsu Y., Near-Band-Edge Optical Responses of Ch$_3$nh$_3$pbcl$_3$ Single Crystals: Photon Recycling of Excitonic Luminescence. *Phys. Rev. Lett.* **2018**, *120*, 057404.

(44)   Ke W.; Stoumpos C. C.; Zhu M.; Mao L.; Spanopoulos I.; Liu J.; Kontsevoi O. Y.; Chen M.; Sarma D.; Zhang Y.; Wasielewski M. R.; Kanatzidis M. G., Enhanced Photovoltaic Performance and Stability with a New Type of Hollow 3d Perovskite Fasni$_3$. *Science Advances* **2017**, *3*.

(45)   Dang Y.; Zhou Y.; Liu X.; Ju D.; Xia S.; Xia H.; Tao X., Formation of Hybrid Perovskite Tin Iodide Single Crystals by Top-Seeded Solution Growth. *Angewandte Chemie International Edition* **2016**, *55*, 3447-3450.

(46)   Tao S. X.; Cao X.; Bobbert P. A., Accurate and Efficient Band Gap Predictions of Metal Halide Perovskites Using the Dft-1/2 Method: Gw Accuracy with Dft Expense. *Sci. Rep.* **2017**, *7*, 14386.

(47)   Veldhuis S. A.; Boix P. P.; Yantara N.; Li M.; Sum T. C.; Mathews N.; Mhaisalkar S. G., Perovskite Materials for Light-Emitting Diodes and Lasers. *Adv. Mater.* **2016**, *28*, 6804-6834.

(48)   Miles R. W.; Zoppi G.; Forbes I., Inorganic Photovoltaic Cells. *Mater Today* **2007**, *10*, 20-27.

(49)   Dong Q.; Fang Y.; Shao Y.; Mulligan P.; Qiu J.; Cao L.; Huang J., Electron-Hole Diffusion Lengths > 175 μm in Solution-Grown CH$_3$NH$_3$PbI$_3$ Single Crystals. *Science* **2015**, *347*, 967-970.

(50)   Shi D.; Adinolfi V.; Comin R.; Yuan M.; Alarousu E.; Buin A.; Chen Y.; Hoogland S.; Rothenberger A.; Katsiev K.; Losovyj Y.; Zhang X.; Dowben P. A.; Mohammed O. F.; Sargent E. H.; Bakr O. M., Low Trap-State Density and Long Carrier Diffusion in Organolead Trihalide Perovskite Single Crystals. *Science* **2015**, *347*, 519-522.

(51)   Møller C. K., Crystal Structure and Photoconductivity of Cæsium Plumbohalides. *Nature* **1958**, *182*, 1436-1436.

(52)   Weber D., *Z. Naturforsch. B* **1978**, *33*, 1443.

(53)   Weber D., *Z. Naturforsch. B* **1978**, *33*, 862.

(54)   Mitzi D. B.; Feild C. A.; Harrison W. T. A.; Guloy A. M., Conducting Tin Halides with a Layered Organic-Based Perovskite Structure. *Nature* **1994**, *369*, 467-469.

(55)   Kagan C. R.; Mitzi D. B.; Dimitrakopoulos C. D., Organic-Inorganic Hybrid Materials as Semiconducting Channels in Thin-Film Field-Effect Transistors. *Science* **1999**, *286*, 945.

(56)   Lee M. M.; Teuscher J.; Miyasaka T.; Murakami T. N.; Snaith H. J., Efficient Hybrid Solar Cells Based on Meso-Superstructured Organometal Halide Perovskites. *Science* **2012**, *338*, 643.

(57)   Tan Z.-K.; Moghaddam R. S.; Lai M. L.; Docampo P.; Higler R.; Deschler F.; Price M.; Sadhanala A.; Pazos L. M.; Credgington D.; Hanusch F.; Bein T.; Snaith H. J.; Friend R. H., Bright Light-Emitting Diodes Based on Organometal Halide Perovskite. *Nat. Nanotechnol.* **2014**, *9*, 687-692.

(58)   Hu X.; Zhang X.; Liang L.; Bao J.; Li S.; Yang W.; Xie Y., High-Performance Flexible Broadband Photodetector Based on Organolead Halide Perovskite. *Adv. Funct. Mater.* **2014**, *24*, 7373-7380.

(59)   Li F.; Ma C.; Wang H.; Hu W.; Yu W.; Sheikh A. D.; Wu T., Ambipolar Solution-Processed Hybrid Perovskite Phototransistors. *Nat. Commun.* **2015**, *6*, 8238.

(60)   Chin X. Y.; Cortecchia D.; Yin J.; Bruno A.; Soci C., Lead Iodide Perovskite Light-Emitting Field-Effect Transistor. *Nat. Commun.* **2015**, *6*, 7383.

(61)   Yoo E. J.; Lyu M.; Yun J.-H.; Kang C. J.; Choi Y. J.; Wang L., Resistive Switching Behavior in Organic–Inorganic Hybrid Ch3nh3pbi3−Xclx Perovskite for Resistive Random Access Memory Devices. *Adv. Mater.* **2015**, *27*, 6170-6175.

(62)   Liao W.-Q.; Zhang Y.; Hu C.-L.; Mao J.-G.; Ye H.-Y.; Li P.-F.; Huang S. D.; Xiong R.-G., A Lead-Halide Perovskite Molecular Ferroelectric Semiconductor. *Nat. Commun.* **2015**, *6*, 7338.

(63)   Xu W.; Cho H.; Kim Y.-H.; Kim Y.-T.; Wolf C.; Park C.-G.; Lee T.-W., Organometal Halide Perovskite Artificial Synapses. *Adv. Mater.* **2016**, *28*, 5916-5922.

(64)   Chen J.-Y.; Chiu Y.-C.; Li Y.-T.; Chueh C.-C.; Chen W.-C., Nonvolatile Perovskite-Based Photomemory with a Multilevel Memory Behavior. *Adv. Mater.* **2017**, *29*, 1702217.

(65)   Wang S.; Liu X.; Li L.; Ji C.; Sun Z.; Wu Z.; Hong M.; Luo J., An Unprecedented Biaxial Trilayered Hybrid Perovskite Ferroelectric with Directionally Tunable Photovoltaic Effects. *J. Am. Chem. Soc.* **2019**, *141*, 7693-7697.

(66)   Chen B.; Rudd P. N.; Yang S.; Yuan Y.; Huang J., Imperfections and Their Passivation in Halide Perovskite Solar Cells. *Chem. Soc. Rev.* **2019**, *48*, 3842-3867.

(67)   Feng J.; Gong C.; Gao H.; Wen W.; Gong Y.; Jiang X.; Zhang B.; Wu Y.; Wu Y.; Fu H.; Jiang L.; Zhang X., Single-Crystalline Layered Metal-Halide Perovskite Nanowires for Ultrasensitive Photodetectors. *Nat. Electron.* **2018**, *1*, 404-410.

(68)   Sánchez F.; Ocal C.; Fontcuberta J., Tailored Surfaces of Perovskite Oxide Substrates for Conducted Growth of Thin Films. *Chem. Soc. Rev.* **2014**, *43*, 2272-2285.

(69)   Hu W. J.; Paudel T. R.; Lopatin S.; Wang Z.; Ma H.; Wu K.; Bera A.; Yuan G.; Gruverman A.; Tsymbal E. Y.; Wu T., Colossal X-Ray-Induced Persistent Photoconductivity in Current-Perpendicular-to-Plane Ferroelectric/Semiconductor Junctions. *Adv. Funct. Mater.* **2018**, *28*, 1704337.

(70)   Jin Hu W.; Wang Z.; Yu W.; Wu T., Optically Controlled Electroresistance and Electrically Controlled Photovoltage in Ferroelectric Tunnel Junctions. *Nat. Commun.* **2016**, *7*, 10808.

(71)   Liu Y.; Weiss N. O.; Duan X.; Cheng H.-C.; Huang Y.; Duan X., Van Der Waals Heterostructures and Devices. *Nat. Rev. Mater.* **2016**, *1*.

(72)   Jariwala D.; Marks T. J.; Hersam M. C., Mixed-Dimensional Van Der Waals Heterostructures. *Nat. Mater.* **2017**, *16*, 170-181.

(73)   Ben Aziza Z.; Henck H.; Pierucci D.; Silly M. G.; Lhuillier E.; Patriarche G.; Sirotti F.; Eddrief M.; Ouerghi A., Van Der Waals Epitaxy of Gase/Graphene Heterostructure: Electronic and Interfacial Properties. *ACS Nano* **2016**, *10*, 9679-9686.

(74)   Zheng W.; Feng W.; Zhang X.; Chen X.; Liu G.; Qiu Y.; Hasan T.; Tan P.; Hu P. A., Anisotropic Growth of Nonlayered Cds on Mos2monolayer for Functional Vertical Heterostructures. *Adv. Funct. Mater.* **2016**, *26*, 2648-2654.

(75)   Wen Y.; Yin L.; He P.; Wang Z.; Zhang X.; Wang Q.; Shifa T. A.; Xu K.; Wang F.; Zhan X.; Wang F.; Jiang C.; He J., Integrated High-Performance Infrared

Phototransistor Arrays Composed of Nonlayered Pbs-Mos2 Heterostructures with Edge Contacts. *Nano Lett.* **2016,** *16,* 6437-6444.

(76)   Chen S.; Teng C.; Zhang M.; Li Y.; Xie D.; Shi G., A Flexible Uv–Vis–Nir Photodetector Based on a Perovskite/Conjugated-Polymer Composite. *Adv. Mater.* **2016,** *28,* 5969-5974.

(77)   He M.; Pang X.; Liu X.; Jiang B.; He Y.; Snaith H.; Lin Z., Monodisperse Dual-Functional Upconversion Nanoparticles Enabled near-Infrared Organolead Halide Perovskite Solar Cells. *Angewandte Chemie International Edition* **2016,** *55,* 4280-4284.

(78)   Ceballos F.; Bellus M. Z.; Chiu H.-Y.; Zhao H., Ultrafast Charge Separation and Indirect Exciton Formation in a Mos2–Mose2 Van Der Waals Heterostructure. *ACS Nano* **2014,** *8,* 12717-12724.

(79)   Georgiou T.; Jalil R.; Belle B. D.; Britnell L.; Gorbachev R. V.; Morozov S. V.; Kim Y. J.; Gholinia A.; Haigh S. J.; Makarovsky O.; Eaves L.; Ponomarenko L. A.; Geim A. K.; Novoselov K. S.; Mishchenko A., Vertical Field-Effect Transistor Based on Graphene-Ws2 Heterostructures for Flexible and Transparent Electronics. *Nat. Nanotechnol.* **2013,** *8,* 100-3.

(80)   Shim J.; Kim H. S.; Shim Y. S.; Kang D. H.; Park H. Y.; Lee J.; Jeon J.; Jung S. J.; Song Y. J.; Jung W. S.; Lee J.; Park S.; Kim J.; Lee S.; Kim Y. H.; Park J. H., Extremely Large Gate Modulation in Vertical Graphene/Wse2 Heterojunction Barristor Based on a Novel Transport Mechanism. *Adv. Mater.* **2016,** *28,* 5293-9.

(81)   Deng Y.; Luo Z.; Conrad N. J.; Liu H.; Gong Y.; Najmaei S.; Ajayan P. M.; Lou J.;   Xu X.; Ye P. D., Black Phosphorus–Monolayer Mos2 Van Der Waals Heterojunction P–N Diode. *ACS Nano* **2014,** *8,* 8292-8299.

(82)   Zhang K.; Zhang T.; Cheng G.; Li T.; Wang S.; Wei W.; Zhou X.; Yu W.; Sun Y.; Wang P.; Zhang D.; Zeng C.; Wang X.; Hu W.; Fan H. J.; Shen G.; Chen X.; Duan X.;   Chang K.; Dai N., Interlayer Transition and Infrared Photodetection in Atomically Thin Type-Ii Mote(2)/Mos(2) Van Der Waals Heterostructures. *ACS Nano* **2016,** *10,* 3852-8.

(83)   Withers F.; Del Pozo-Zamudio O.; Mishchenko A.; Rooney A. P.; Gholinia A.; Watanabe K.; Taniguchi T.; Haigh S. J.; Geim A. K.; Tartakovskii A. I.; Novoselov K. S., Light-Emitting Diodes by Band-Structure Engineering in Van Der Waals Heterostructures. *Nat. Mater.* **2015,** *14,* 301-6.

(84)   Wang Q.; Wen Y.; He P.; Yin L.; Wang Z.; Wang F.; Xu K.; Huang Y.; Wang F.;   Jiang C.; He J., High-Performance Phototransistor of Epitaxial Pbs Nanoplate-Graphene Heterostructure with Edge Contact. *Adv. Mater.* **2016,** *28,* 6497-503.

(85)   Ruzmetov D.; Zhang K.; Stan G.; Kalanyan B.; Bhimanapati G. R.; Eichfeld S. M.; Burke R. A.; Shah P. B.; O'regan T. P.; Crowne F. J.; Birdwell A. G.; Robinson J.

A.; Davydov A. V.; Ivanov T. G., Vertical 2d/3d Semiconductor Heterostructures Based on Epitaxial Molybdenum Disulfide and Gallium Nitride. *ACS Nano* **2016**, *10*, 3580-8.

(86) Yang T.; Wang X.; Zheng B.; Qi Z.; Ma C.; Fu Y.; Fu Y.; Hautzinger M. P.; Jiang Y.; Li Z.; Fan P.; Li F.; Zheng W.; Luo Z.; Liu J.; Yang B.; Chen S.; Li D.; Zhang L.; Jin S.; Pan A., Ultrahigh-Performance Optoelectronics Demonstrated in Ultrathin Perovskite-Based Vertical Semiconductor Heterostructures. *ACS Nano* **2019**, *13*, 7996-8003.

(87) Yang W.; Chen J.; Zhang Y.; Zhang Y.; He J.-H.; Fang X., Silicon-Compatible Photodetectors: Trends to Monolithically Integrate Photosensors with Chip Technology. *Adv. Funct. Mater.* **2019**, *29*, 1808182.

(88) Lopez-Sanchez O.; Alarcon Llado E.; Koman V.; Fontcuberta I Morral A.; Radenovic A.; Kis A., Light Generation and Harvesting in a Van Der Waals Heterostructure. *ACS Nano* **2014**, *8*, 3042-3048.

(89) Yang T.; Zheng B.; Wang Z.; Xu T.; Pan C.; Zou J.; Zhang X.; Qi Z.; Liu H.; Feng Y.; Hu W.; Miao F.; Sun L.; Duan X.; Pan A., Van Der Waals Epitaxial Growth and Optoelectronics of Large-Scale Wse2/Sns2 Vertical Bilayer P–N Junctions. *Nat. Commun.* **2017**, *8*, 1906.

(90) Niu L.; Liu X.; Cong C.; Wu C.; Wu D.; Chang T. R.; Wang H.; Zeng Q.; Zhou J.; Wang X.; Fu W.; Yu P.; Fu Q.; Najmaei S.; Zhang Z.; Yakobson B. I.; Tay B. K.; Zhou W.; Jeng H. T.; Lin H.; Sum T. C.; Jin C.; He H.; Yu T.; Liu Z., Controlled Synthesis of Organic/Inorganic Van Der Waals Solid for Tunable Light-Matter Interactions. *Adv. Mater.* **2015**, *27*, 7800-8.

(91) Wang Y.; Gao L.; Yang Y.; Xiang Y.; Chen Z.; Dong Y.; Zhou H.; Cai Z.; Wang G.-C.; Shi J., Nontrivial Strength of Van Der Waals Epitaxial Interaction in Soft Perovskites. *Phys. Rev. Mater.* **2018**, *2*, 076002.

(92) Zheng W.; Feng W.; Zhang X.; Chen X. S.; Liu G. B.; Qiu Y. F.; Hasan T.; Tan P. H.; Hu P. A., Anisotropic Growth of Nonlayered Cds on Mos2 Monolayer for Functional Vertical Heterostructures. *Adv. Funct. Mater.* **2016**, *26*, 2648-2654.

(93) Zakharchenko K. V.; Katsnelson M. I.; Fasolino A., Finite Temperature Lattice Properties of Graphene Beyond the Quasiharmonic Approximation. *Phys. Rev. Lett.* **2009**, *102*, 046808.

(94) Das A.; Pisana S.; Chakraborty B.; Piscanec S.; Saha S. K.; Waghmare U. V.; Novoselov K. S.; Krishnamurthy H. R.; Geim A. K.; Ferrari A. C.; Sood A. K., Monitoring Dopants by Raman Scattering in an Electrochemically Top-Gated Graphene Transistor. *Nat. Nanotechnol.* **2008**, *3*, 210-5.

(95)   Jnawali G.; Rao Y.; Beck J. H.; Petrone N.; Kymissis I.; Hone J.; Heinz T. F., Observation of Ground- and Excited-State Charge Transfer at the C60/Graphene Interface. *ACS Nano* **2015**, *9*, 7175-7185.

(96)   Ni Z. H.; Yu T.; Lu Y. H.; Wang Y. Y.; Feng Y. P.; Shen Z. X., Uniaxial Strain on Graphene: Raman Spectroscopy Study and Band-Gap Opening (Vol 2, Pg 2301, 2008). *ACS Nano* **2009**, *3*, 483-483.

(97)   Bera A.; Peng H. Y.; Lourembam J.; Shen Y. D.; Sun X. W.; Wu T., A Versatile Light-Switchable Nanorod Memory: Wurtzite Zno on Perovskite Srtio3. *Adv. Funct. Mater.* **2013**, *23*, 4977-4984.

(98)   Liu C. H.; Chang Y. C.; Norris T. B.; Zhong Z., Graphene Photodetectors with Ultra-Broadband and High Responsivity at Room Temperature. *Nat. Nanotechnol.* **2014**, *9*, 273-8.

(99)   Yu X.; Li Y.; Hu X.; Zhang D.; Tao Y.; Liu Z.; He Y.; Haque M. A.; Liu Z.; Wu T.; Wang Q. J., Narrow Bandgap Oxide Nanoparticles Coupled with Graphene for High Performance Mid-Infrared Photodetection. *Nat. Commun.* **2018**, *9*, 4299.

(100)   Chen Y.; Yi H. T.; Wu X.; Haroldson R.; Gartstein Y. N.; Rodionov Y. I.; Tikhonov K. S.; Zakhidov A.; Zhu X. Y.; Podzorov V., Extended Carrier Lifetimes and Diffusion in Hybrid Perovskites Revealed by Hall Effect and Photoconductivity Measurements. *Nat. Commun.* **2016**, *7*, 12253.

(101)   Galkowski K.; Mitioglu A.; Miyata A.; Plochocka P.; Portugall O.; Eperon G. E.; Wang J. T. W.; Stergiopoulos T.; Stranks S. D.; Snaith H. J.; Nicholas R. J., Determination of the Exciton Binding Energy and Effective Masses for Methylammonium and Formamidinium Lead Tri-Halide Perovskite Semiconductors. *Energy & Environmental Science* **2016**, *9*, 962-970.

(102)   Kim S.; Menabde S. G.; Jang M. S., Efficient Photodoping of Graphene in Perovskite–Graphene Heterostructure. *Adv. Electron. Mater.* **2019**.

(103)   Hoye R. L. Z.; Lai M.-L.; Anaya M.; Tong Y.; Gałkowski K.; Doherty T.; Li W.; Huq T. N.; Mackowski S.; Polavarapu L.; Feldmann J.; Macmanus-Driscoll J. L.; Friend R. H.; Urban A. S.; Stranks S. D., Identifying and Reducing Interfacial Losses to Enhance Color-Pure Electroluminescence in Blue-Emitting Perovskite Nanoplatelet Light-Emitting Diodes. *ACS Energy Letters* **2019**, *4*, 1181-1188.

(104)   Dursun I.; Zheng Y.; Guo T.; De Bastiani M.; Turedi B.; Sinatra L.; Haque M. A.; Sun B.; Zhumekenov A. A.; Saidaminov M. I.; García De Arquer F. P.; Sargent E. H.; Wu T.; Gartstein Y. N.; Bakr O. M.; Mohammed O. F.; Malko A. V., Efficient Photon Recycling and Radiation Trapping in Cesium Lead Halide Perovskite Waveguides. *ACS Energy Letters* **2018**, *3*, 1492-1498.

(105) Zhang Y.; Sun R.; Ou X.; Fu K.; Chen Q.; Ding Y.; Xu L.-J.; Liu L.; Han Y.; Malko A. V.; Liu X.; Yang H.; Bakr O. M.; Liu H.; Mohammed O. F., Metal Halide Perovskite Nanosheet for X-Ray High-Resolution Scintillation Imaging Screens. *ACS Nano* **2019**, *13*, 2520-2525.

(106) Kim Y. C.; Kim K. H.; Son D.-Y.; Jeong D.-N.; Seo J.-Y.; Choi Y. S.; Han I. T.; Lee S. Y.; Park N.-G., Printable Organometallic Perovskite Enables Large-Area, Low-Dose X-Ray Imaging. *Nature* **2017**, *550*, 87-91.

(107) Hu W.; Huang W.; Yang S.; Wang X.; Jiang Z.; Zhu X.; Zhou H.; Liu H.; Zhang Q.; Zhuang X.; Yang J.; Kim D. H.; Pan A., High-Performance Flexible Photodetectors Based on High-Quality Perovskite Thin Films by a Vapor–Solution Method. *Adv. Mater.* **2017**, *29*, 1703256.

(108) Ran J.; Ma T. Y.; Gao G.; Du X.-W.; Qiao S. Z., Porous P-Doped Graphitic Carbon Nitride Nanosheets for Synergistically Enhanced Visible-Light Photocatalytic H2 Production. *Energy & Environmental Science* **2015**, *8*, 3708-3717.

(109) Velusamy D. B.; Haque M. A.; Parida M. R.; Zhang F.; Wu T.; Mohammed O. F.; Alshareef H. N., 2d Organic–Inorganic Hybrid Thin Films for Flexible Uv–Visible Photodetectors. *Adv. Funct. Mater.* **2017**, *27*, 1605554.

(110) Niu W.; Yang Y., Graphitic Carbon Nitride for Electrochemical Energy Conversion and Storage. *ACS Energy Letters* **2018**, *3*, 2796-2815.

(111) Guo S.; Deng Z.; Li M.; Jiang B.; Tian C.; Pan Q.; Fu H., Phosphorus-Doped Carbon Nitride Tubes with a Layered Micro-Nanostructure for Enhanced Visible-Light Photocatalytic Hydrogen Evolution. *Angew. Chem. Int. Ed. Engl.* **2016**, *55*, 1830-4.

(112) Zhu Y.-P.; Ren T.-Z.; Yuan Z.-Y., Mesoporous Phosphorus-Doped G-C3n4 Nanostructured Flowers with Superior Photocatalytic Hydrogen Evolution Performance. *ACS Appl. Mater. Interfaces* **2015**, *7*, 16850-16856.

(113) Yaghoubi H.; Li Z.; Chen Y.; Ngo H. T.; Bhethanabotla V. R.; Joseph B.; Ma S.; Schlaf R.; Takshi A., Toward a Visible Light-Driven Photocatalyst: The Effect of Midgap-States-Induced Energy Gap of Undoped Tio2 Nanoparticles. *ACS Catalysis* **2015**, *5*, 327-335.

(114) Zheng G.; Zhu C.; Ma J.; Zhang X.; Tang G.; Li R.; Chen Y.; Li L.; Hu J.; Hong J.; Chen Q.; Gao X.; Zhou H., Manipulation of Facet Orientation in Hybrid Perovskite Polycrystalline Films by Cation Cascade. *Nat. Commun.* **2018**, *9*, 2793.

(115) Shao S.; Dong J.; Duim H.; Ten Brink G. H.; Blake G. R.; Portale G.; Loi M. A., Enhancing the Crystallinity and Perfecting the Orientation of Formamidinium Tin Iodide for Highly Efficient Sn-Based Perovskite Solar Cells. *Nano Energy* **2019**, *60*, 810-816.

(116) Ke W.; Fang G.; Wan J.; Tao H.; Liu Q.; Xiong L.; Qin P.; Wang J.; Lei H.; Yang G.; Qin M.; Zhao X.; Yan Y., Efficient Hole-Blocking Layer-Free Planar Halide Perovskite Thin-Film Solar Cells. *Nat. Commun.* **2015,** *6,* 6700.

(117) Lee Y.; Kwon J.; Hwang E.; Ra C.-H.; Yoo W. J.; Ahn J.-H.; Park J. H.; Cho J. H., High-Performance Perovskite–Graphene Hybrid Photodetector. *Adv. Mater.* **2015,** *27,* 41-46.

(118) Muduli S.; Pandey P.; Devatha G.; Babar R.; M T.; Kothari D. C.; Kabir M.; Pillai P. P.; Ogale S., Photoluminescence Quenching in Self-Assembled Cspbbr3 Quantum Dots on Few-Layer Black Phosphorus Sheets. *Angewandte Chemie International Edition* **2018,** *57,* 7682-7686.

(119) Im J. H.; Jang I. H.; Pellet N.; Gratzel M.; Park N. G., Growth of Ch3nh3pbi3 Cuboids with Controlled Size for High-Efficiency Perovskite Solar Cells. *Nat. Nanotechnol.* **2014,** *9,* 927-32.

(120) Saidaminov M. I.; Adinolfi V.; Comin R.; Abdelhady A. L.; Peng W.; Dursun I.; Yuan M.; Hoogland S.; Sargent E. H.; Bakr O. M., Planar-Integrated Single-Crystalline Perovskite Photodetectors. *Nat. Commun.* **2015,** *6,* 8724.

(121) Gu L.; Tavakoli M. M.; Zhang D.; Zhang Q.; Waleed A.; Xiao Y.; Tsui K.-H.; Lin Y.; Liao L.; Wang J.; Fan Z., 3d Arrays of 1024-Pixel Image Sensors Based on Lead Halide Perovskite Nanowires. *Adv. Mater.* **2016,** *28,* 9713-9721.

(122) Deng W.; Zhang X.; Huang L.; Xu X.; Wang L.; Wang J.; Shang Q.; Lee S.-T.; Jie J., Aligned Single-Crystalline Perovskite Microwire Arrays for High-Performance Flexible Image Sensors with Long-Term Stability. *Adv. Mater.* **2016,** *28,* 2201-2208.

(123) Han Q.; Bae S.-H.; Sun P.; Hsieh Y.-T.; Yang Y.; Rim Y. S.; Zhao H.; Chen Q.; Shi W.; Li G.; Yang Y., Single Crystal Formamidinium Lead Iodide (Fapbi3): Insight into the Structural, Optical, and Electrical Properties. *Adv. Mater.* **2016,** *28,* 2253-2258.

(124) Wang Y.; Fullon R.; Acerce M.; Petoukhoff C. E.; Yang J.; Chen C.; Du S.; Lai S. K.; Lau S. P.; Voiry D.; O'carroll D.; Gupta G.; Mohite A. D.; Zhang S.; Zhou H.; Chhowalla M., Solution-Processed Mos2/Organolead Trihalide Perovskite Photodetectors. *Adv. Mater.* **2017,** *29,* 1603995.

(125) Zhang H.; Liao Q.; Wang X.; Hu K.; Yao J.; Fu H., Controlled Substitution of Chlorine for Iodine in Single-Crystal Nanofibers of Mixed Perovskite Mapbi3–Xclx. *Small* **2016,** *12,* 3780-3787.

(126) Ji L.; Hsu H.-Y.; Lee J. C.; Bard A. J.; Yu E. T., High-Performance Photodetectors Based on Solution-Processed Epitaxial Grown Hybrid Halide Perovskites. *Nano Lett.* **2018,** *18,* 994-1000.

(127)   Wang F.; Mei J.; Wang Y.; Zhang L.; Zhao H.; Zhao D., Fast Photoconductive Responses in Organometal Halide Perovskite Photodetectors. *ACS Appl. Mater. Interfaces* **2016,** *8,* 2840-2846.

(128)   Gao L.; Zeng K.; Guo J.; Ge C.; Du J.; Zhao Y.; Chen C.; Deng H.; He Y.; Song H.;   Niu G.; Tang J., Passivated Single-Crystalline Ch3nh3pbi3 Nanowire Photodetector with High Detectivity and Polarization Sensitivity. *Nano Lett.* **2016,** *16,* 7446-7454.

(129)   Guo Y.; Liu C.; Tanaka H.; Nakamura E., Air-Stable and Solution-Processable Perovskite Photodetectors for Solar-Blind Uv and Visible Light. *The Journal of Physical Chemistry Letters* **2015,** *6,* 535-539.

(130)   Endres J.; Egger D. A.; Kulbak M.; Kerner R. A.; Zhao L.; Silver S. H.; Hodes G.; Rand B. P.; Cahen D.; Kronik L.; Kahn A., Valence and Conduction Band Densities of States of Metal Halide Perovskites: A Combined Experimental-Theoretical Study. *J Phys Chem Lett* **2016,** *7,* 2722-9.

(131)   Zhang Y.; Mori T.; Niu L.; Ye J., Non-Covalent Doping of Graphitic Carbon Nitride Polymer with Graphene: Controlled Electronic Structure and Enhanced Optoelectronic Conversion. *Energy & Environmental Science* **2011,** *4,* 4517-4521.

(132)   Song J.; Xu L.; Li J.; Xue J.; Dong Y.; Li X.; Zeng H., Monolayer and Few-Layer All-Inorganic Perovskites as a New Family of Two-Dimensional Semiconductors for Printable Optoelectronic Devices. *Adv. Mater.* **2016,** *28,* 4861-9.

(133)   Deng H.; Yang X.; Dong D.; Li B.; Yang D.; Yuan S.; Qiao K.; Cheng Y. B.; Tang J.; Song H., Flexible and Semitransparent Organolead Triiodide Perovskite Network Photodetector Arrays with High Stability. *Nano Lett.* **2015,** *15,* 7963-9.

(134)   Liu C.; Wang K.; Yi C.; Shi X.; Du P.; Smith A. W.; Karim A.; Gong X., Ultrasensitive Solution-Processed Perovskite Hybrid Photodetectors. *J. Mater. Chem. C* **2015,** *3,* 6600-6606.

(135)   Guo Y.; Liu C.; Tanaka H.; Nakamura E., Air-Stable and Solution-Processable Perovskite Photodetectors for Solar-Blind Uv and Visible Light. *J Phys Chem Lett* **2015,** *6,* 535-9.

(136)   Wang Q.; Chen B.; Liu Y.; Deng Y.; Bai Y.; Dong Q.; Huang J., Scaling Behavior of Moisture-Induced Grain Degradation in Polycrystalline Hybrid Perovskite Thin Films. *Energy & Environmental Science* **2017,** *10,* 516-522.

(137)   Yuan Y.; Huang J., Ion Migration in Organometal Trihalide Perovskite and Its Impact on Photovoltaic Efficiency and Stability. *Acc. Chem. Res.* **2016,** *49,* 286-293.

(138)   Kim H.;   Wang Z.; Alshareef H. N., Mxetronics: Electronic and Photonic Applications of Mxenes. *Nano Energy* **2019,** *60,* 179-197.

(139)   Hantanasirisakul K.; Gogotsi Y., Electronic and Optical Properties of 2d Transition Metal Carbides and Nitrides (Mxenes). *Adv. Mater.* **2018,** *30,* 1804779.

(140)   Li J.; Levitt A.; Kurra N.; Juan K.; Noriega N.; Xiao X.; Wang X.; Wang H.; Alshareef H. N.; Gogotsi Y., Mxene-Conducting Polymer Electrochromic Microsupercapacitors. *Energy Storage Materials* **2019,** *20,* 455-461.

(141)   Zhang C.; Mckeon L.; Kremer M. P.; Park S.-H.; Ronan O.; Seral⬜Ascaso A.; Barwich S.; Coileáin C. Ó.; Mcevoy N.; Nerl H. C.; Anasori B.; Coleman J. N.; Gogotsi Y.; Nicolosi V., Additive-Free Mxene Inks and Direct Printing of Micro-Supercapacitors. *Nat. Commun.* **2019,** *10,* 1795.

(142)   Ming F.; Liang H.; Huang G.; Bayhan Z.; Alshareef H. N., Mxenes for Rechargeable Batteries Beyond the Lithium-Ion. *Adv. Mater.* **2021,** *33,* 2004039.

(143)   Wang D.; Li F.; Lian R.; Xu J.; Kan D.; Liu Y.; Chen G.; Gogotsi Y.; Wei Y., A General Atomic Surface Modification Strategy for Improving Anchoring and Electrocatalysis Behavior of $Ti_3c_2t_2$ Mxene in Lithium–Sulfur Batteries. *ACS Nano* **2019,** *13,* 11078-11086.

(144)   Shahzad F.; Alhabeb M.; Hatter C. B.; Anasori B.; Man Hong S.; Koo C. M.; Gogotsi Y., Electromagnetic Interference Shielding with 2d Transition Metal Carbides (Mxenes). *Science* **2016,** *353,* 1137.

(145)   Velusamy D. B.; El-Demellawi J. K.; El-Zohry A. M.; Giugni A.; Lopatin S.; Hedhili M. N.; Mansour A. E.; Fabrizio E. D.; Mohammed O. F.; Alshareef H. N., Mxenes for Plasmonic Photodetection. *Adv. Mater.* **2019,** *31,* 1807658.

(146)   Wang Z.; Kim H.; Alshareef H. N., Oxide Thin-Film Electronics Using All-Mxene Electrical Contacts. *Adv. Mater.* **2018,** *30,* 1706656.

(147)   Yang L.; Dall'agnese C.; Dall'agnese Y.; Chen G.; Gao Y.; Sanehira Y.; Jena A. K.; Wang X.-F.; Gogotsi Y.; Miyasaka T., Surface-Modified Metallic $Ti_3c_2t_x$ Mxene as Electron Transport Layer for Planar Heterojunction Perovskite Solar Cells. *Adv. Funct. Mater.* **2019,** *29,* 1905694.

(148)   Xu H.; Ren A.; Wu J.; Wang Z., Recent Advances in 2d Mxenes for Photodetection. *Adv. Funct. Mater.* **2020,** *30,* 2000907.

(149)   Ahn S.; Han T.-H.; Maleski K.; Song J.; Kim Y.-H.; Park M.-H.; Zhou H.; Yoo S.; Gogotsi Y.; Lee T.-W., A 2d Titanium Carbide Mxene Flexible Electrode for High-Efficiency Light-Emitting Diodes. *Adv. Mater.* **2020,** *32,* 2000919.

(150)   Yang Y.; Jeon J.; Park J.-H.; Jeong M. S.; Lee B. H.; Hwang E.; Lee S., Plasmonic Transition Metal Carbide Electrodes for High-Performance Inse Photodetectors. *ACS Nano* **2019,** *13,* 8804-8810.

(151)   Maleski K.; Shuck C. E.; Fafarman A. T.; Gogotsi Y., The Broad Chromatic Range of Two-Dimensional Transition Metal Carbides. *Advanced Optical Materials* **2021**, *9*, 2001563.

(152)   Dillon A. D.; Ghidiu M. J.; Krick A. L.; Griggs J.; May S. J.; Gogotsi Y.; Barsoum M. W.; Fafarman A. T., Highly Conductive Optical Quality Solution-Processed Films of 2d Titanium Carbide. *Adv. Funct. Mater.* **2016**, *26*, 4162-4168.

(153)   Dorodnyy A.; Salamin Y.; Ma P.; Plestina J. V.; Lassaline N.; Mikulik D.; Romero-Gomez P.; Morral A. F. I.; Leuthold J., Plasmonic Photodetectors. *IEEE J. Sel. Top. Quantum Electron.* **2018**, *24*, 1-13.

(154)   Ratchford D. C., Plasmon-Induced Charge Transfer: Challenges and Outlook. *ACS Nano* **2019**, *13*, 13610-13614.

(155)   Goykhman I.; Desiatov B.; Khurgin J.; Shappir J.; Levy U., Waveguide Based Compact Silicon Schottky Photodetector with Enhanced Responsivity in the Telecom Spectral Band. *Opt. Express* **2012**, *20*, 28594-28602.

(156)   Grajower M.; Desiatov B.; Mazurski N.; Shappir J.; Khurgin J. B.; Levy U., Optimization and Experimental Demonstration of Plasmonic Enhanced Internal Photoemission Silicon Schottky Detectors in the Mid-Ir. *ACS Photonics* **2017**, *4*, 1015-1020.

(157)   Zhang C.; Beidaghi M.; Naguib M.; Lukatskaya M. R.; Zhao M.-Q.; Dyatkin B.; Cook K. M.; Kim S. J.; Eng B.; Xiao X.; Long D.; Qiao W.; Dunn B.; Gogotsi Y., Synthesis and Charge Storage Properties of Hierarchical Niobium Pentoxide/Carbon/Niobium Carbide (Mxene) Hybrid Materials. *Chem. Mater.* **2016**, *28*, 3937-3943.

(158)   Kamysbayev V.; Filatov A. S.; Hu H.; Rui X.; Lagunas F.; Wang D.; Klie R. F.; Talapin D. V., Covalent Surface Modifications and Superconductivity of Two-Dimensional Metal Carbide Mxenes. *Science* **2020**, *369*, 979.

(159)   El-Demellawi J. K.; Lopatin S.; Yin J.; Mohammed O. F.; Alshareef H. N., Tunable Multipolar Surface Plasmons in 2d Ti3c2tx Mxene Flakes. *ACS Nano* **2018**, *12*, 8485-8493.

(160)   Shi Z.; Zhang Y.; Cui C.; Li B.; Zhou W.; Ning Z.; Mi Q., Symmetrization of the Crystal Lattice of Mapbi3 Boosts the Performance and Stability of Metal–Perovskite Photodiodes. *Adv. Mater.* **2017**, *29*, 1701656.

(161)   Bao C.; Chen Z.; Fang Y.; Wei H.; Deng Y.; Xiao X.; Li L.; Huang J., Low-Noise and Large-Linear-Dynamic-Range Photodetectors Based on Hybrid-Perovskite Thin-Single-Crystals. *Adv. Mater.* **2017**, *29*, 1703209.

(162) Ma Z.; Zhang Y.; Li T.; Tang X.; Zhao H.; Li J.; Ma C.; Yao J., High-Performance Self-Powered Perovskite Photodetector for Visible Light Communication. *Applied Physics A* **2020**, *126*, 869.

(163) Zhou H.; Zeng J.; Song Z.; Grice C. R.; Chen C.; Song Z.; Zhao D.; Wang H.; Yan Y., Self-Powered All-Inorganic Perovskite Microcrystal Photodetectors with High Detectivity. *The Journal of Physical Chemistry Letters* **2018**, *9*, 2043-2048.

(164) Zhang S.; Jiao H.; Wang X.; Chen Y.; Wang H.; Zhu L.; Jiang W.; Liu J.; Sun L.; Lin T.; Shen H.; Hu W.; Meng X.; Pan D.; Wang J.; Zhao J.; Chu J., Highly Sensitive Insb Nanosheets Infrared Photodetector Passivated by Ferroelectric Polymer. *Adv. Funct. Mater.* **2020**, *30*, 2006156.

(165) Bian Z.; Tachikawa T.; Zhang P.; Fujitsuka M.; Majima T., Au/Tio2 Superstructure-Based Plasmonic Photocatalysts Exhibiting Efficient Charge Separation and Unprecedented Activity. *J. Am. Chem. Soc.* **2014**, *136*, 458-465.

(166) Ng C.; Cadusch J. J.; Dligatch S.; Roberts A.; Davis T. J.; Mulvaney P.; Gómez D. E., Hot Carrier Extraction with Plasmonic Broadband Absorbers. *ACS Nano* **2016**, *10*, 4704-4711.

(167) Tagliabue G.; Jermyn A. S.; Sundararaman R.; Welch A. J.; Duchene J. S.; Pala R.; Davoyan A. R.; Narang P.; Atwater H. A., Quantifying the Role of Surface Plasmon Excitation and Hot Carrier Transport in Plasmonic Devices. *Nat. Commun.* **2018**, *9*, 3394.

(168) Liu L.; Fang W.-H.; Long R.; Prezhdo O. V., Lewis Base Passivation of Hybrid Halide Perovskites Slows Electron–Hole Recombination: Time-Domain Ab Initio Analysis. *The Journal of Physical Chemistry Letters* **2018**, *9*, 1164-1171.

(169) Noel N. K.; Abate A.; Stranks S. D.; Parrott E. S.; Burlakov V. M.; Goriely A.; Snaith H. J., Enhanced Photoluminescence and Solar Cell Performance Via Lewis Base Passivation of Organic–Inorganic Lead Halide Perovskites. *ACS Nano* **2014**, *8*, 9815-9821.

(170) Lin Y.; Shen L.; Dai J.; Deng Y.; Wu Y.; Bai Y.; Zheng X.; Wang J.; Fang Y.; Wei H.; Ma W.; Zeng X. C.; Zhan X.; Huang J., Π-Conjugated Lewis Base: Efficient Trap-Passivation and Charge-Extraction for Hybrid Perovskite Solar Cells. *Adv. Mater.* **2017**, *29*, 1604545.

(171) Wang F.; Geng W.; Zhou Y.; Fang H.-H.; Tong C.-J.; Loi M. A.; Liu L.-M.; Zhao N., Phenylalkylamine Passivation of Organolead Halide Perovskites Enabling High-Efficiency and Air-Stable Photovoltaic Cells. *Adv. Mater.* **2016**, *28*, 9986-9992.

(172) Chattopadhyay P.; Raychaudhuri B., Frequency Dependence of Forward Capacitance-Voltage Characteristics of Schottky Barrier Diodes. *Solid-State Electron.* **1993**, *36*, 605-610.

(173) Padma R.; Lakshmi B. P.; Reddy V. R., Capacitance–Frequency (C–F) and Conductance–Frequency (G–F) Characteristics of Ir/N-Ingan Schottky Diode as a Function of Temperature. *Superlattices Microstruct.* **2013**, *60*, 358-369.

(174) Doğan H.; Yıldırım N.; Orak İ.; Elagöz S.; Turut A., Capacitance-Conductance-Frequency Characteristics of Au/Ni/N-Gan/Undoped Gan Structures. *Physica B: Condensed Matter* **2015**, *457*, 48-53.

(175) Van Mensfoort S. L. M.; Coehoorn R., Determination of Injection Barriers in Organic Semiconductor Devices from Capacitance Measurements. *Phys. Rev. Lett.* **2008**, *100*, 086802.

(176) Zeng W.; Liu X.; Guo X.; Niu Q.; Yi J.; Xia R.; Min Y., Morphology Analysis and Optimization: Crucial Factor Determining the Performance of Perovskite Solar Cells. *Molecules* **2017**, *22*.

(177) Kong W.; Rahimi-Iman A.; Bi G.; Dai X.; Wu H., Oxygen Intercalation Induced by Photocatalysis on the Surface of Hybrid Lead Halide Perovskites. *The Journal of Physical Chemistry C* **2016**, *120*, 7606-7611.

(178) Pérez-Osorio M. A.; Lin Q.; Phillips R. T.; Milot R. L.; Herz L. M.; Johnston M. B.; Giustino F., Raman Spectrum of the Organic–Inorganic Halide Perovskite Ch3nh3pbi3 from First Principles and High-Resolution Low-Temperature Raman Measurements. *The Journal of Physical Chemistry C* **2018**, *122*, 21703-21717.

(179) Burgio L.; Clark R. J. H.; Firth S., Raman Spectroscopy as a Means for the Identification of Plattnerite (Pbo), of Lead Pigments and of Their Degradation Products. *Analyst* **2001**, *126*, 222-227.

(180) Peng W.; Wang L.; Murali B.; Ho K.-T.; Bera A.; Cho N.; Kang C.-F.; Burlakov V. M.; Pan J.; Sinatra L.; Ma C.; Xu W.; Shi D.; Alarousu E.; Goriely A.; He J.-H.; Mohammed O. F.; Wu T.; Bakr O. M., Solution-Grown Monocrystalline Hybrid Perovskite Films for Hole-Transporter-Free Solar Cells. *Adv. Mater.* **2016**, *28*, 3383-3390.

(181) Chen J.; Morrow D. J.; Fu Y.; Zheng W.; Zhao Y.; Dang L.; Stolt M. J.; Kohler D. D.; Wang X.; Czech K. J.; Hautzinger M. P.; Shen S.; Guo L.; Pan A.; Wright J. C.; Jin S., Single-Crystal Thin Films of Cesium Lead Bromide Perovskite Epitaxially Grown on Metal Oxide Perovskite (Srtio3). *J. Am. Chem. Soc.* **2017**, *139*, 13525-13532.

(182) Liu P.; He X.; Ren J.; Liao Q.; Yao J.; Fu H., Organic–Inorganic Hybrid Perovskite Nanowire Laser Arrays. *ACS Nano* **2017**, *11*, 5766-5773.

(183) Wang G.; Li D.; Cheng H. C.; Li Y.; Chen C. Y.; Yin A.; Zhao Z.; Lin Z.; Wu H.; He Q.; Ding M.; Liu Y.; Huang Y.; Duan X., Wafer-Scale Growth of Large Arrays of Perovskite Microplate Crystals for Functional Electronics and Optoelectronics. *Sci Adv* **2015**, *1*, e1500613.

(184) Lin C.-H.; Cheng B.; Li T.-Y.; Retamal J. R. D.; Wei T.-C.; Fu H.-C.; Fang X.; He J.-H., Orthogonal Lithography for Halide Perovskite Optoelectronic Nanodevices. *ACS Nano* **2019**, *13*, 1168-1176.

(185) Liu Z.; Alshareef H. N., Mxenes for Optoelectronic Devices. *Adv. Electron. Mater.* **2021**, *n/a*, 2100295.

(186) Gao S.; Liu G.; Yang H.; Hu C.; Chen Q.; Gong G.; Xue W.; Yi X.; Shang J.; Li R.-W., An Oxide Schottky Junction Artificial Optoelectronic Synapse. *ACS Nano* **2019**, *13*, 2634-2642.

(187) Wang Y.; Yin L.; Huang W.; Li Y.; Huang S.; Zhu Y.; Yang D.; Pi X., Optoelectronic Synaptic Devices for Neuromorphic Computing. *Advanced Intelligent Systems* **2021**, *3*, 2000099.

(188) Sirringhaus H.; Tessler N.; Friend R. H., Integrated Optoelectronic Devices Based on Conjugated Polymers. *Science* **1998**, *280*, 1741.

(189) Chen S.; Lou Z.; Chen D.; Shen G., An Artificial Flexible Visual Memory System Based on an Uv-Motivated Memristor. *Adv. Mater.* **2018**, *30*, 1705400.

www.ingramcontent.com/pod-product-compliance
Lightning Source LLC
LaVergne TN
LVHW042159190726

843493LV00006B/1738